ESSENTIAL MATHEMATICS
for
GENERAL CHEMISTRY

Errata Sheet
for
ESSENTIAL MATHEMATICS FOR GENERAL CHEMISTRY
by
Robert L. Osburn

Page 22 –

Examples; under (a):

Correct answer should be: $= 10.$ (correct answer)

Page 58 –

Examples; under (c) should be: $(0.00052)^2 = ?$

Page 81 –

Last line of (h) should be: $= 2 \times 10^{-10}$

Page 93 –

Quadratic formula should be: $x = \dfrac{-b \pm \sqrt{b^2 - 4ac}}{2a}$

Word "Examples" should be: "Example"

Correct equation under Example: $x^2 + (4 \times 10^{-4})x - (4 \times 10^{-5}) = 0$

$$x = 1 \qquad \text{should be:} \qquad a = 1$$
$$x = 4 \times 10^{-4} \qquad\qquad\qquad b = 4 \times 10^{-4}$$
$$x = -4 \times 10^{-5} \qquad\qquad\qquad c = -4 \times 10^{-5}$$

Quadratic formula at bottom of page should be:

$$x = \frac{-4 \times 10^{-4} \pm \sqrt{(4 \times 10^{-4})^2 - 4(-4 \times 10^{-5})}}{2}$$

Page 98 –

Quadratic formula in middle of page should be:

$$x = \frac{-b \pm \sqrt{b^2 - 4ac}}{2a}$$

ESSENTIAL MATHEMATICS
for
GENERAL CHEMISTRY

Robert L. Osburn, Ph.D.

JOHN WILEY & SONS, INC.
New York • London • Sydney • Toronto

L.C. 74-32574

ISBN 0 471-65704-2

Printed in the United States of America

10 9 8 7 6 5 4 3 2 1

iv

PREFACE

This book will help students to survive general chemistry by removing certain arithmetical obstacles so that more time can be devoted to chemical principles. I think that "little things" such as powers, roots, unit conversions, and logarithms are, too often, the downfall of beginning students. Somehow, in training even mathematically competent students, these topics are not covered or are not given proper emphasis. I believe that most students (particularly the ones with marginal backgrounds) who flounder in general chemistry do so because of arithmetic, not chemical concepts. This volume will be a life preserver for these students.

I am grateful to Miss Muffet for typing the original manuscript.

Robert L. Osburn, Ph. D.

TABLE OF CONTENTS

Unit 1

EXPONENTS;
POWERS OF TEN

In solving problems in chemistry and the sciences in general, it is often necessary to deal with very large and very small numbers. For instance, in one gram of the element hydrogen there are:

602,300,000,000,000,000,000,000

atoms of hydrogen. Therefore, the mass of <u>one</u> hydrogen atom is:

0.000,000,000,000,000,000,000,00166 grams

Numbers such as these are both awkward and unwieldy to work with. It is easier to express these numbers using <u>exponential notation,</u> that is, a number between one and ten (the coefficient) multiplied by a power of ten (the exponential). Some common powers of ten are given in Table 1-1.

To express a number in exponential notation, the following form is used:

$$N = 10^{\text{exponent}}$$

where by convention N is a number between 1 and 10 (the coefficient) and the exponent is a whole number, either positive or negative. To determine the exponent, count the number of places that the decimal point must be moved in order to give N, the coefficient. <u>If the decimal point must be moved to the left, the exponent is a positive whole number. If the decimal point must be moved to the right, the exponent is a negative whole number.</u>

TABLE 1-1

$$1 \ = \ 1 \ \text{x} \ 10^0$$

$$10 \ = \ 1 \ \text{x} \ 10^1$$

$$100 \ = \ 1 \ \text{x} \ 10^2 \ \text{i.e.,} \ (1) \ (10) \ (10)$$

$$1000 \ = \ 1 \ \text{x} \ 10^3 \ \text{i.e.,} \ (1) \ (10) \ (10) \ (10)$$

$$10,000 \ = \ 1 \ \text{x} \ 10^4 \ \text{i.e.,} \ (1) \ (10) \ (10) \ (10) \ (10)$$

$$100,000 \ = \ 1 \ \text{x} \ 10^5 \ \text{i.e.,} \ (1) \ (10) \ (10) \ (10) \ (10) \ (10)$$

$$1,000,000 \ = \ 1 \ \text{x} \ 10^6 \ \text{i.e.,} \ (1) \ (10) \ (10) \ (10) \ (10) \ (10) \ (10)$$

$$0.1 = 1 \ \text{x} \ 10^{-1} \ \text{i.e.,} \ \frac{1}{1 \ \text{x} \ 10^1}$$

$$0.01 = 1 \ \text{x} \ 10^{-2} \ \text{i.e.,} \ \frac{1}{1 \ \text{x} \ 10^2}$$

$$0.001 = 1 \ \text{x} \ 10^{-3} \ \text{i.e.,} \ \frac{1}{1 \ \text{x} \ 10^3}$$

$$0.0001 = 1 \ \text{x} \ 10^{-4} \ \text{i.e.,} \ \frac{1}{1 \ \text{x} \ 10^4}$$

$$0.00001 = 1 \ \text{x} \ 10^{-5} \ \text{i.e.,} \ \frac{1}{1 \ \text{x} \ 10^5}$$

$$0.000001 = 1 \ \text{x} \ 10^{-6} \ \text{i.e.,} \ \frac{1}{1 \ \text{x} \ 10^6}$$

Examples:

$$1,230,000,000 = 1.23 \times 10^9$$
(significant to 3 places; to be discussed in Unit 2)

Coefficient   Power of Ten
(between 1 and 10)

and

$$602,300,000,000,000,000,000 = 6.023 \times 10^{23}$$

and

$$0.000,000,000,36 = 3.6 \times 10^{-10}$$

and

$$0.000,000,000,000,000,000,000,00166 = 1.66 \times 10^{-24}$$

ADDITION AND SUBTRACTION OF EXPONENTIAL NUMBERS

In order to add or subtract exponential numbers, each number <u>must</u> be <u>expressed</u> to the <u>same power.</u> The coefficients are then added or sub-tracted in the convential manner, and the powers of ten remain <u>unaltered.</u>

Examples:

Carry out the indicated operations.

(a) $(5.00 \times 10^{-5}) + (3.00 \times 10^{-3}) + (4.00 \times 10^{-4})$

$$5.00 \times 10^{-5} = 0.05 \times 10^{-3}$$

and

$$4.00 \times 10^{-4} = 0.40 \times 10^{-3}$$

therefore:

$$(0.05 \times 10^{-3}) + (3.00 \times 10^{-3}) + (0.40 \times 10^{-3}) = 3.45 \times 10^{-3}$$

(b) $(5.0 \times 10^{-6}) - (4.0 \times 10^{-7})$

       $4.0 \times 10^{-7} = 0.4 \times 10^{-6}$

therefore:

$$(5.0 \times 10^{-6}) - (0.4 \times 10^{-6}) = 4.6 \times 10^{-6}$$

MULTIPLICATION OF EXPONENTIAL NUMBERS

In multiplying very large and very small numbers, first express the numbers in standard exponential notation. The coefficients are then multiplied in the usual manner (i.e., ordinary multiplication) and the powers of ten (exponents) are added, algebraically, in accordance with the multiplication law for exponents.

Examples:

(a) $(5.0 \times 10^{4})(1.5 \times 10^{2}) = ?$

       $(5.0)(1.5) \times 10^{4} \times 10^{2} = 7.5 \times 10^{4+2}$

                      $= 7.5 \times 10^{6}$

(b) $(4.2 \times 10^{-8})(2.0 \times 10^{3}) = ?$

       $(4.2)(2.0) \times 10^{-8} \times 10^{3} = 8.4 \times 10^{-8+3}$

                      $= 8.4 \times 10^{-5}$

(c) $(5,000)(2,000,000) = ?$

       $(5 \times 10^{3})(2 \times 10^{6}) \quad =$

$$(5)(2) \times 10^3 \times 10^6 = 10 \times 10^{3+6}$$
$$= 10 \times 10^9$$
$$= 1 \times 10^{10}$$

(Note that in the final answer the coefficient is expressed as a number between 1 and 10, i.e. standard exponential notation.)

(d) $(0.000,000,2)(4,000) = ?$

$$(2 \times 10^{-7})(4 \times 10^3) =$$

$$(2)(4) \times 10^{-7} \times 10^3 = 8 \times 10^{-7+3}$$
$$= 8 \times 10^{-4}$$

DIVISION OF EXPONENTIAL NUMBERS

In dividing very large and very small numbers, the numbers should be first expressed in standard exponential notation. The coefficients are then handled by ordinary division and the exponent in the <u>denominator</u> is <u>subtracted from</u> the exponent in the numerator, <u>algebraically</u>.

Examples:

(a) $\dfrac{5.0 \times 10^4}{2.0 \times 10^2} = ?$

$$\frac{5.0}{2.0} \times \frac{10^4}{10^2} = 2.5 \times 10^{4-2}$$

$$= 2.5 \times 10^2$$

(b) $\dfrac{6.0 \times 10^{-3}}{4.0 \times 10^{6}}$ = ?

$\dfrac{6.0}{4.0} \times \dfrac{10^{-3}}{10^{6}}$ = $1.5 \times 10^{-3-6}$

$= 1.5 \times 10^{-9}$

(c) $\dfrac{3.6 \times 10^{-5}}{6 \times 10^{-4}}$ = ?

$\dfrac{3.6}{6} \times \dfrac{10^{-5}}{10^{-4}}$ = $0.6 \times 10^{-5-(-4)}$

$= 0.6 \times 10^{-5+4}$

$= 0.6 \times 10^{-1}$
(Not standard exponential notation.)

$= 6 \times 10^{-2}$

POWERS AND ROOTS OF EXPONENTIAL NUMBERS

In both raising an exponential number to a power (i.e. squaring, cubing, etc.) and extracting a root of an exponential number (i.e., square root, cube root, etc.), the same general principle applies, i.e.:

$$(10^{a})^{b} = 10^{ab}$$

The difference between the two operations lies in the fact that in the case of raising an exponential number to a power, "b" is a whole number (i.e., 2, 3, etc.), whereas in extracting a root of an exponential number, "b" is a fractional power (i.e., 1/2, 1/3, etc.). Stated differently, when extracting roots of exponential numbers, the general relationship may be written as:

$$\sqrt[n]{10^{a}} = (10^{a})^{1/n}$$

Squaring of Exponential Numbers

In squaring exponential numbers, the coefficients are multiplied in the usual manner, and the <u>exponent</u> is <u>multiplied by two.</u>

Examples:

(a) $(3.0 \times 10^2)^2 = ?$

$(3.0)^2 \times (10^2)^2 =$

$(3.0)(3.0) \times 10^2 \times 10^2 = 9.0 \times 10^4$

(b) $(4.0 \times 10^{-6})^2 = ?$

$(4.0)(4.0) \times 10^{-6} \times 10^{-6} = 16 \times 10^{-12}$

$= 1.6 \times 10^{-11}$

(c) $(0.000,05)^2 = ?$

$(5 \times 10^{-5})^2 = 25 \times 10^{-10}$

$= 2.5 \times 10^{-9}$

Cubing of Exponential Numbers

In cubing exponential numbers, the coefficients are cubed in the usual manner, and the <u>exponent</u> is <u>multiplied by three.</u>

Examples:

(a) $(2 \times 10^3)^3 = ?$

$(2)^3 \times (10^3)^3 = (2)(2)(2) \times 10^3 \times 10^3 \times 10^3$

$$= 8 \times 10^9$$

(b) $(4 \times 10^{-2})^3 = ?$

$$(4)(4)(4) \times 10^{-2} \times 10^{-2} \times 10^{-2} = 64 \times 10^{-6}$$

$$= 6.4 \times 10^{-5}$$

(c) $(30,000)^3 = ?$

$$(3 \times 10^4)^3 = (3)(3)(3) \times 10^{12}$$

$$= 27 \times 10^{12}$$

$$= 2.7 \times 10^{13}$$

Square Roots of Exponential Numbers

Remember, that in extracting roots of exponential numbers the general relationship,

$$(10^a)^b = 10^{ab}$$

still applies. The difference here, however, is that now "b" is a <u>fractional power</u>, i.e., 1/2. The general relationship,

$$\sqrt[n]{10^a} = (10^a)^{1/n}$$

now becomes,

$$\sqrt[2]{10^a} = (10^a)^{1/2}$$

Therefore, the exponential number must be written in such a way so that the <u>power of ten</u> is <u>divisible by 2</u>, to give a whole number.

Examples:

(a) $\sqrt{9 \times 10^4}$ $= (9 \times 10^4)^{1/2}$

$\qquad\qquad = (9)^{1/2} \times (10^4)^{1/2}$

$\qquad\qquad = 3 \times 10^2$

(b) $\sqrt{1.6 \times 10^{-7}}$ $= (1.6 \times 10^{-7})^{1/2}$

Note that here, the <u>power of 10</u> is <u>not divisble by 2</u> to give a whole number. Therefore, the exponential number must be expressed in such a way so that the power of 10 <u>is</u> divisible by 2.

Since, $1.6 \times 10^{-7} = 16 \times 10^{-8}$

therefore,

$\qquad (16 \times 10^{-8})^{1/2} = 4 \times 10^{-4}$

Cube Roots of Exponential Numbers

The same general principle applies here as above with extracting square roots of exponential numbers, except that in extracting cube roots the exponential number must be written in such a way so that the <u>power of ten is divisible by 3,</u> to give a whole number.

Examples:

(a) $\sqrt[3]{64 \times 10^6}$ $= (64 \times 10^6)^{1/3}$

$\qquad\qquad = (64)^{1/3} \times (10^6)^{1/3}$

$\qquad\qquad = 4 \times 10^2$

(b) $\sqrt[3]{2.7 \times 10^{-8}} \quad = (2.7 \times 10^{-8})^{1/3}$

Here, the exponent is <u>not</u> divisible by 3, to give a whole number.

Therefore,

$$(2.7 \times 10^{-8})^{1/3} \quad = (27 \times 10^{-9})^{1/3}$$

$$= (27)^{1/3} \times (10^{-9})^{1/3}$$

$$= 3 \times 10^{-3}$$

(c) $(8 \times 10^{3})^{2/3} \quad = \left[(8 \times 10^{3})^{2} \right]^{1/3}$

$$= (64 \times 10^{6})^{1/3}$$

$$= 4 \times 10^{2}$$

In rewriting exponential numbers so that the power of 10 will be divisible by 2 (for square roots) or 3 (for cube roots), the exponential number should be written in such a way so that the coefficient will always have one, two, or three digits to the left of the decimal place, i.e., the coefficient should never be less than one. The reason for following this convention will be – come apparent when the use of the slide rule for extracting roots is discuss– ed in Unit 4.

PROBLEMS

1.   Express each of the following in exponential notation.

a.	7,200,000	j.	0.273
b.	22,400	k.	0.00000021
c.	22.4	l.	10.6
d.	8,000	m.	830×10^{4}
e.	125,000	n.	72×10^{-5}
f.	1.4	o.	0.0035×10^{-1}
g.	0.007	p.	1973×10^{-5}
h.	0.00035	q.	0.0002×10^{6}
i.	0.00905	r.	0.5×10^{2}

2.   Express each of the following in arithmetical form.

a.	7×10^{-3}	e.	9.3×10^{-10}
b.	2.1×10^{5}	f.	6×10^{23}
c.	1.4×10^{0}	g.	1.7×10^{-24}
d.	5.05×10^{-4}	h.	2.1×10^{1}

3.   Perform the following additions and subtractions.

a. $4.5 \times 10^{4} + 1.6 \times 10^{4}$

b. $3.1 \times 10^{-3} + 1.4 \times 10^{-3}$

c. $6.1 \times 10^{2} + 8.2 \times 10^{2}$

d. $6.5 \times 10^{-2} - 1.8 \times 10^{-2}$

e. $2.4 \times 10^{5} - 1.9 \times 10^{5}$

f. $8.27 \times 10^5 - 1.40 \times 10^4$

g. $1.43 \times 10^{-7} + 1.80 \times 10^{-8}$

h. $9.55 \times 10^4 + 8.00 \times 10^2$

i. $9.73 \times 10^{-10} - 7.3 \times 10^{-11}$

j. $5.00 \times 10^{-4} + 3.00 \times 10^{-2} - 4.00 \times 10^{-3}$

k. $6.00 \times 10^{-4} + 5.00 \times 10^{-2} - 5.00 \times 10^{-3}$

l. $8.00 \times 10^{-3} + 7.00 \times 10^{-2} - 9 \times 10^{-4}$

4.   Perform the following multiplications. Express all answers in standard exponential notation.

(a) $(3 \times 10^2)(4 \times 10^3)$

(b) $(1.2 \times 10^4)(1.6 \times 10^{-3})$

(c) $(8.0 \times 10^{-1})(4 .2 \times 10^{-3})$

(d) $(3.0 \times 10^5)(1.5 \times 10^5)$

(e) $(9.8 \times 10^{-4})(1.6 \times 10^{-3})$

(f) $(1.5 \times 10^{-10})(2.6 \times 10^8)$

(g) $(1.9 \times 10^{25})(1.3 \times 10^{13})$

(h) $(3 \times 10^4)(4 \times 10^2)(6 \times 10^5)$

(i) $(4 \times 10^{-3})(3 \times 10^{-2})(5 \times 10^{-5})$

(j) $(1.5 \times 10^{-3})(2.0 \times 10^4)(1.7 \times 10^{-8})$

(k) $(4.2 \times 10^6)(1.8 \times 10^5)(2.1 \times 10^{-3})$

(l) $(9.1 \times 10^{-7})(8.4 \times 10^{-4})(3.6 \times 10^{-5})$

(m) $(3.5 \times 10^2)(1.2 \times 10^4)(6.4 \times 10^3)(1.7 \times 10^8)$

(n) $(2.1 \times 10^{-1})(4.0 \times 10^{-3})(2.5 \times 10^{-6})(7.4 \times 10^{-2})$

(o) $(4.7 \times 10^{4})(9.1 \times 10^{-2})(8.6 \times 10^{-3})(1.3 \times 10^{5})$

5.   Perform the following divisions. Express all answers in standard exponential notation.

(a) $\dfrac{3 \times 10^{4}}{2 \times 10^{3}}$ (h) $\dfrac{8.4 \times 10^{-12}}{7.0 \times 10^{10}}$

(b) $\dfrac{6 \times 10^{5}}{2 \times 10^{3}}$ (i) $\dfrac{9.5 \times 10^{14}}{1.5 \times 10^{-14}}$

(c) $\dfrac{7.4 \times 10^{-5}}{2.1 \times 10^{-3}}$ (j) $\dfrac{8.2 \times 10^{-15}}{4.1 \times 10^{-20}}$

(d) $\dfrac{9.3 \times 10^{-6}}{3.1 \times 10^{-4}}$ (k) $\dfrac{6.5 \times 10^{5}}{1.2 \times 10^{20}}$

(e) $\dfrac{1.5 \times 10^{6}}{3.0 \times 10^{5}}$ (l) $\dfrac{1.4 \times 10^{15}}{7.0 \times 10^{12}}$

(f) $\dfrac{1.7 \times 10^{-5}}{2.6 \times 10^{4}}$ (m) $\dfrac{7.29 \times 10^{20}}{1.46 \times 10^{-5}}$

(g) $\dfrac{4.58 \times 10^{2}}{2.29 \times 10^{4}}$ (n) $\dfrac{9.5 \times 10^{23}}{8.6 \times 10^{22}}$

6.   Perform the indicated operations. Express all answers in standard exponential notation.

(a) $\dfrac{(5.0 \times 10^{-5}) - (5.0 \times 10^{-6})}{9.0 \times 10^{3}}$

(b) $\dfrac{9.0 \times 10^{4}}{(5.0 \times 10^{-6}) - (5.0 \times 10^{-7})}$

(c) $\dfrac{8.2 \times 10^4}{(4.0 \times 10^{-6}) + (5.0 \times 10^{-7})}$

(d) $\dfrac{(7.2 \times 10^4) + (8.0 \times 10^3)}{(6.1 \times 10^{-5})}$

(e) $\dfrac{(1.5 \times 10^{-3}) \, (2.6 \times 10^{-5})}{(4.1 \times 10^{-6})}$

(f) $\dfrac{(8.4 \times 10^5) \, (1.7 \times 10^6)}{(3.4 \times 10^4)}$

(g) $\dfrac{(1.7 \times 10^{-3}) \, (2.4 \times 10^4)}{(1.9 \times 10^{-2}) \, (2.1 \times 10^{-5})}$

(h) $\dfrac{(9.1 \times 10^7) \, (8.2 \times 10^{-3})}{(5.5 \times 10^{-4}) \, (2.7 \times 10^{-7})}$

(i) $\dfrac{(1.2 \times 10^2) \, (1.8 \times 10^5) \, (2.3 \times 10^{-4})}{(8.1 \times 10^{-2}) \, (1.7 \times 10^6) \, (3.2 \times 10^{-5})}$

7.   Perform the indicated operations; first write each number in exponential form. Express all answers in standard exponential notation.

(a)   (25,000) (1,600)

(b)   (2,500,000) (1,700,000)

(c)   (180) (230) (14,000)

(d)   (8,300) (19,500) (100,000)

(e)   (0.009) (0.00005)

(f)   (0.0000065) (0.00000021)

(g)   (0.08) (0.00042) (0.000018)

(h)   (0.000075) (0.00000029) (0.0000000017)

(i)   (28,000) (0.00016)

(j)   (0.0000064) (120,000)

(k)   (0.0048) (0.000065) (1,350,000)

(l)   (2,400) (14,500) (0.0073)

(m) $\dfrac{(68,500)\ (0.000027)}{(1,700)\ (0.00083)}$

(n) $\dfrac{(0.0000085)\ (0.0000023)}{(0.000041)\ (0.0000052)}$

(o) $\dfrac{(8,550,000)\ (12,300)\ (175)}{(0.00035)\ (0.0000064)}$

(p) $\dfrac{(0.00000028)\ (0.0000017)\ (0.0045)}{(17,800)\ (5,280)\ (273)}$

8.   Perform the indicated operations with powers and roots. Express all answers in standard exponential notation.

(a) $(2 \times 10^4)^2$

(b) $(4.1 \times 10^3)^2$

(c) $(6.3 \times 10^{-2})^2$

(d) $(1.2 \times 10^{-8})^2$

(e) $(9.4 \times 10^9)^2$

(f) $(1.4 \times 10^2)^3$

(g) $(1.9 \times 10^{-2})^3$

(h) $(2.4 \times 10^{-5})^3$

(i) $(4.0 \times 10^{14})^3$

(j) $(2.0 \times 10^{-20})^3$

(k) $(0.0016)^2$

(l) $(0.027)^3$

(m) $(50,000)^3$

(n) $(500,000)^2$

(o) $(0.0000006)^2$

(p) $(0.00000003)^3$

(q) $\dfrac{1}{(48,000)^2}$

(r) $\dfrac{17.5}{(0.000035)^2}$

(s) $\dfrac{150}{(0.000005)^3}$

(t) $\dfrac{1}{(18,000)^3}$

(u) $(0.0016)^{1/2}$

(v) $(1 \times 10^{14})^{1/2}$

(w) $(1.21 \times 10^{-14})^{1/2}$

(x) $(1.25 \times 10^{-4})^{1/3}$

(y) $(8.1 \times 10^{11})^{1/2}$

(z) $(6.4 \times 10^{7})^{1/3}$

9.   Perform the indicated operations. Express all answers in standard
      exponential notation.

   (a) $(4 \times 10^4)^2 \times (1 \times 10^{12})^{1/2}$

   (b) $(2.1 \times 10^{-3})^2 \times (1.5 \times 10^2)^3$

   (c) $(1.44 \times 10^4)^{1/2} \times (6.4 \times 10^7)^{1/3}$

   (d) $(0.0064)^{1/2} \times (1.25 \times 10^5)^{1/3}$

   (e) $\dfrac{(3.0 \times 10^{10})^2}{(1.0 \times 10^5)^3}$

   (f) $\dfrac{(2.5 \times 10^{-3})^3}{(1.5 \times 10^4)^2}$

   (g) $\left[\dfrac{1.25 \times 10^{10}}{5 \times 10^4} \right]^{1/2}$

   (h) $\left[\dfrac{1.28 \times 10^{-4}}{2.0 \times 10^3} \right]^{1/3}$

Unit 2

SIGNIFICANT FIGURES

A significant figure is a figure which is known to be reasonably reliable. The numerical value of every observed measurement is always an approximation, because no physical measurement (e.g. mass, length, volume, time) is _ever_ exactly correct due to limitations on the reliability of the measuring instrument. As an example, assume that the length of an object has been recorded as 18.7 cm. This means that the length was measured to the _nearest_ tenth of a cm, i.e., there is uncertainty in the seven. The _exact_ value is somewhere between 18.75 and 18.65. The value 18.7, therefore, has _three_ significant figures. This same measurement recorded as 18.70 cm represents _four_ significant figures, i.e., here the uncertainty is understood to be plus or minus one unit in the second decimal place.

When zeros are present in numbers, confusion often arises when determining the correct number of significant figures. Whether zeros at the end of a number are considered as significant or not in many cases must be decided on the basis of common sense. If the population of city is given as 8,875,000 people, there are _probably_ only 4 significant figures due to the uncertainty of present census-gathering techniques.

Sometimes a decimal point is used to indicate the number of significant figures in numbers ending in zero. In the quantity, "3800 people", the number of significant figures could be two, three, or four. However, when this quantity is written as "3800. people", four significant figures are indicated. A more common way of writing this quantity is to use exponential notation, i.e., 3.800×10^3 (four significant figures).

A final consideration concerning zeros in relation to significant figures is necessary. Zeros which appear as the first figures of a number are _not_ significant, because they are there _only_ to locate the decimal point, i.e., 0.045 has _two_ significant figures. This number is written as 4.5×10^{-2} in exponential notation. Table 2-1 illustrates the above points.

18

ROUNDING OFF SIGNIFICANT FIGURES

A number is rounded off to the desired number of significant figures by dropping one or more digits to the right. The following rules should be observed when rounding off numbers.

(1)   When the first digit dropped is less than 5, the last digit retained remains unchanged.

That is,

3.273 rounded to 3 significant figures becomes 3.27

(2)   When the first digit dropped is more than 5, the last digit retained is increased by 1.

That is,

3.486 rounded to 3 significant figures becomes 3.49

(3)   When the first digit dropped is exactly 5, the last digit retained is increased by 1 if it is odd and remains unchanged if it is even.

That is,

142.75 rounded to 4 significant figures becomes 142.8

and,

142,65 rounded to 4 significant figures becomes 142.6

MATHEMATICAL OPERATIONS INVOLVING SIGNIFICANT FIGURES

Addition and Subtraction

When inexact numbers are added or subtracted, the sum or difference cannot be more precise than the least precise number involved in the cal-

TABLE 2-1

Quantity	Number of Significant Figures
10.53 grams	4
842 pounds	3
0.005860 meters	4 (note that the last zero is a significant figure)
1.5×10^4 miles	2
$875,754.05	8
725,000 people	3 (probably)
725,000. people	6
7.25×10^5 people	3
7.250×10^5 people	4

culation. It is important to realize that in adding or subtracting inexact numbers that the number of significant figures in the answer is <u>not</u> governed by the quantity having the fewest significant figures used in the calculation.

Examples:

Add the following quantities.

(a) 35.450 g   (5 significant figures)
      5.625 g   (4 significant figures)
     0.127 g   (3 significant figures)

     41.202 g   (5 significant figures)

Note that here 5 significant figures are justified in the answer because each value involved in the calculation has a precision of + or − 0.001 g (± 0.001 g), i.e., the uncertainty is in the third decimal place for all three quantities.

(b) 75.1   g
      0.0025   g
     0.00004 g

     75.10254 g = 75.1 g (correct answer)

Since the first number is actually 75.1 ± 0.1 g, the sum cannot be more precise than ± 0.1 g.

(c) 5.40   g
      2.7752 g
     0.023   g

     8.1982   g = 8.20 g (correct answer)

Another method that is sometimes used is to round off the individual numbers before performing the arithmetic, retaining only as many numbers to the right of the decimal as there are in the least precise number. Using this method, example (c) above becomes:

            5.40   g
            2.78   g
           0.02   g
            8.20   g

Subtract the following.

(d)   255.75
       −16.5

       239.25 = 239.2 (correct answer)

Using the alternate method this same problem becomes:

                        255.8
                        −16.5
                        239.3

Note that this answer differs by one in the last place from the previous answer. Remember, however, that the last place is <u>known</u> to have some uncertainty in it (i.e., ± 0.1).

Multiplication and Division

In the multiplication and division of inexact numbers, the answer must be rounded off so that it will contain only as many significant figures as are contained in the least precise number involved in the calculation.

Examples:

Carry out the following operations.

(a)   (4.25) (2.4) = 10.2

                    = 10   (correct answer)

(b)   (4.10) (1.630) (2.0000) = 13.366

                               = 13.4   (correct answer)

(c)   (1.75 x 10^5) (2.1 x 10^6) = 3.675 x 10^{11}

                                = 3.7 x 10^{11} (correct answer)

(d) $(200,000.) (1.5) = 300,000$

$\qquad\qquad\qquad = 3.0 \times 10^5$ (correct answer)

(e) $\dfrac{3.2}{1.5043} = \dfrac{3.2}{1.5} = 2.133$

$\qquad\qquad\qquad = 2.1$ (correct answer)

(f) $\dfrac{183.5}{7.175} = 25.58$ (correct answer)

(g) $\dfrac{1.58 \times 10^{-7}}{2,000,000.} = \dfrac{1.58 \times 10^{-7}}{2.00 \times 10^6}$

$\qquad\qquad = 0.79 \times 10^{-13}$

$\qquad\qquad = 7.90 \times 10^{-14}$ (correct answer)

PROBLEMS

1.   Determine the correct number of significant figures in each of the following.

    (a)   415 meters

    (b)   0.0028 gram

    (c)   0.0530 kilogram

    (d)   1.7 pounds

    (e)   0.10040 ton

    (f)   0.125 mole

    (g) 6×10^{23} atoms

    (h) 6.023×10^{23} molecules

    (i)   143,217.4 meters

    (j)   1.0028 milliliters

    (k)   10.4728 grams

    (l) 2.96×10^{-4} mole

    (m)   0.0000001 meter

    (n) 1×10^{8} centimeters

2.   Perform the indicated operations, rounding off the answer to the correct number of significant figures.

    (a)   Add   28.45
                             5.246
                           148.2

    (b)   Subtract   355.88
                          − 21.313

    (c)   (0.09385) + (1.072613) + (3.1925)

    (d) $(8.25 \times 10^{5}) - (1.5 \times 10^{4})$

(e) $(1.45 \times 10^{-7}) + (2.80 \times 10^{-8})$

(f) $(3.0 \times 10^5) (1.5 \times 10^5)$

(g) $(0.038) (146) (1.5 \times 10^5)$

(h) $\dfrac{1.8 \times 10^{-6}}{2.1 \times 10^5}$

(i) $\dfrac{7.35 \times 10^{15}}{1.26 \times 10^{-5}}$

(j) $\dfrac{9.2 \times 10^{-20}}{4.36 \times 10^{-14}}$

(k) $\dfrac{(8.5 \times 10^5) (1.8 \times 10^6)}{(3.4 \times 10^8)}$

(l) $\dfrac{(9.13 \times 10^{-2}) (1.526 \times 10^{-4})}{(5.1 \times 10^3) (2.463 \times 10^{10})}$

(m) $\dfrac{(9,000,000) (15,400) (173)}{(0.00265) (0.0075)}$

(n) $(2 \times 10^{-3})^2$

(o) $(9.4 \times 10^9)^2$

(p) $(2.5 \times 10^{-5})^3$

(q) $(0.00015)^2$

(r) $\dfrac{14.2}{(0.00025)^2}$

(s) $(2.162)^2$

(t) $(1.7852)^3$

(u) $(9.00 \times 10^6)^{1/2}$

(v) $(1.6 \times 10^{-7})^{1/2}$

(w) $(64 \times 10^9)^{1/3}$

(x) $(2.7 \times 10^{-8})^{1/3}$

Unit 3

THE METRIC SYSTEM;

CONVERSION FACTORS

METRIC SYSTEM

The metric system which is the International System of Measurement, is used exclusively in chemistry as well as other branches of the sciences. An advantage that this system has over the more familiar English system is that the metric system is based on multiples of 10 while no such common base exists in the English system. In the metric system, the fundamental dimensions are length, mass, and time are measured by the meter, the gram, and the second respectively. Volume, in the metric system, is measured in liters which is derived from the fundamental unit of length.

Scientific measurements range from extremely large to extremely small numbers and units that are appropriate for one measurement may not be appropriate for another. For this reason, it is common practice to vary the size of a fundamental unit by adding a suitable prefix to it. Common metric prefixes and their indicated values are given in Table 3-1. The most commonly used of these prefixes and their relationships to the three fundamental units in the metric system are shown in Table 3-2.

CONVERSION FACTORS

It is possible to express the same quantity in a number of different ways depending upon the units which are used to express the quantity, i.e., the statement that a certain object is one foot long is the same as saying that the object is 12 inches long. These two statements are completely equivalent, the only difference is that different units have been used to express the quantity, i.e.

TABLE 3-1

METRIC PREFIXES

Prefix	Symbol	Value
tera	T	10^{12}
giga	G	10^{9}
mega	M	10^{6}
kilo	k	10^{3}
hecto	h	10^{2}
deka	da	10^{1}
deci	d	10^{-1}
centi	c	10^{-2}
milli	m	10^{-3}
micro	μ	10^{-6}
nano	n	10^{-9}
pico	p	10^{-12}
femto	f	10^{-15}
atto	a	10^{-18}

TABLE 3-2

---

1 meter (m) $= 10^1$ decimeters (dm) 1 dm $= 10^{-1}$ m

$\qquad\qquad\;\; = 10^2$ decimeters (cm) 1 cm $= 10^{-2}$ m

$\qquad\qquad\;\; = 10^3$ millimeters (mm) 1 mm $= 10^{-3}$ m

$\qquad\qquad\;\; = 10^6$ micrometers (μm) 1 μm $= 10^{-6}$ m

$\qquad\qquad\quad$ (or micron, μ)   (1 $\mu = 10^{-6}$ m)

$\qquad\qquad\;\; = 10^{-3}$ kilometer (km) 1 km $= 10^3$ m

1 gram (g) $= 10^1$ decigrams (dg) 1 dg $= 10^{-1}$ g

$\qquad\qquad\;\; = 10^2$ centigrams (cg) 1 cg $= 10^{-2}$ g

$\qquad\qquad\;\; = 10^3$ milligrams (mg) 1 mg $= 10^{-3}$ g

$\qquad\qquad\;\; = 10^6$ micrograms (μg) 1μg $= 10^{-6}$ g

$\qquad\qquad\;\; = 10^{-3}$ kilogram (kg) 1 kg $= 10^3$ g

1 second (sec)$= 10^3$ milliseconds (msec) 1 msec $= 10^{-3}$ sec

$\qquad\qquad\;\; = 10^6$ microseconds (μsec) 1 sec $= 10^{-6}$ sec

1 liter (l) $= 10^3$ milliliters (ml) 1 ml $= 10^{-3}$ l

$\qquad\qquad\;\; = 10^3$ cubic centimeters 1 cm$^3 = 10^{-3}$ l

$\qquad\qquad\quad$ (cm^3 or cc)

$\qquad\qquad\;\; = 10^6$ microliters (μl) 1μl $= 10^{-6}$ l

---

1 foot = 12 inches

This simple relationship is the basis for the use of all conversion factors. Notice that this conversion factor can be written in two ways, both of which are completely equivalent, i.e.,

$$\frac{1 \text{ ft}}{12 \text{ in}} \quad \text{or} \quad \frac{12 \text{ in}}{1 \text{ ft}}$$

Written in one of these forms, this statement of equivalency can be used to either convert feet to inches or inches to feet, since

$$\frac{1 \text{ ft}}{12 \text{ in}} = \frac{12 \text{ in}}{1 \text{ ft}} = 1,$$

and multiplication of any quantity by one does not change its value. That is, the magnitude of a quantity is not altered by a conversion of the units of that quantity.

Convert:

3 ft to inches.

$$3 \text{ ft} \times \frac{12 \text{ in}}{1 \text{ ft}} = 36 \text{ in.}$$

or convert

96 inches to feet

$$96 \text{ in} \times \frac{1 \text{ ft}}{12 \text{ in}} = 8 \text{ ft}$$

Note that the form of the conversion factor which is used depends upon the direction of the conversion, i.e., inches to feet or feet to inches. When the conversion is from feet to inches the conversion factor is used which will allow cancellation of the units of feet leaving only units of inches, i.e., $\frac{12 \text{ in}}{1 \text{ ft}}$. For the conversion of inches to feet, the other form of the conversion factor is the one which must be used. Some commonly used conversion factors are given in Table 3-3.

TABLE 3-3

CONVERSION FACTORS

1 meter (m)	= 39.37 in
1 km	= 1000 m
1 mile	= 1.61 km
1 mile	= 5,280 ft
1 yard	= 0.9144 m
1 foot	= 30.5 cm
1 inch	= 2.54 cm
1 angstrom (Å)	= 10^{-8} cm
1 liter	= 1.0567 qt (U.S.)
1 gal (U.S.)	= 3.7854 l
1 ft^3	= 28.3 l
1 cm^3	= 1 ml
1 lb	= 453.59 g
1 lb	= 16 oz
1 oz	= 28.35 g
1 ton	= 2000 lb (U.S.)
1 ton (metric)	= 1000 kg

These conversion factors are used for expressing equivalent quantities in different ways, i.e., using different units. Conversion factors are used for making a conversion of units from the English system to the Metric system, or vice versa, or for making a conversion of units within the same system. Several examples will illustrate the use of conversion factors.

Examples:

(a)   Convert 5 g to decigrams.

1 g = 10 dg

therefore, the two forms of this equivalent statement are:

$\dfrac{1 \text{ g}}{10 \text{ dg}}$ and $\dfrac{10 \text{ dg}}{1 \text{ g}}$

In order to convert g to dg, the conversion factor must be used which will cancel out the units of grams leaving only units of decigrams, i.e., $\dfrac{10 \text{ dg}}{1 \text{ g}}$

therefore,

$$5 \text{ g} \times \frac{10 \text{ dg}}{1 \text{ g}} = 50 \text{ dg}$$

(b)   Convert 75 dm to meters.

Using the relationship, 1 m = 10 dm

$$75 \text{ dm} \times \frac{1 \text{ m}}{10 \text{ dm}} = 7.5 \text{ m}$$

(c)   Convert 3.50 ml to liters.

Since, $1 \text{ l} = 10^{3} \text{ ml}$

$$3.50 \text{ ml} \times \frac{1 \text{ l}}{10^{3} \text{ ml}} = 3.50 \times 10^{-3} \text{ l}$$

(d)   Convert 17.25 kg to milligrams.

Using the relationships, $1 \text{ kg} = 10^3$ g and $1 \text{ mg} = 10^{-3}$ g,

$$17.25 \text{ kg} \times \frac{10^3 \text{ g}}{1 \text{ kg}} \times \frac{1 \text{ mg}}{10^{-3} \text{ g}} = 17,250,000 \text{ mg}$$

$$= 1.725 \times 10^7 \text{ mg (in exponential form)}$$

Note that in problems like this where a series of conversions is required, the entire problem should first be set up <u>before</u> any calculations are carried out. A check of the cancellation of units can then be made before any arithmetical operations are done in order to insure that the conversion factors are set up in the correct form, i.e., only when all units <u>except</u> the units of the answer cancel.

(e)   Given that the diameter of an atom of gold (Au) is 3 A, what is its diameter in inches?

The conversion factors needed are:

$$1 \text{ A} = 10^{-8} \text{ cm}$$

and,

$$1 \text{ in} = 2.54 \text{ cm}$$

therefore,

$$3 \text{ A} \times \frac{10^{-8} \text{ cm}}{1 \text{ A}} \times \frac{1 \text{ in}}{2.54 \text{ cm}} = 1.18 \times 10^{-8} \text{ in}$$

(f)   Express the volume of 10 gallons using Metric system units.

Conversion factors needed:

1 liter = 1.06 quart

1 gallon = 4 quarts

therefore,

$$10 \text{ gal} \times \frac{4 \text{ qt}}{1 \text{ gal}} \times \frac{1 \text{ l}}{1.06 \text{ qt}} = 37.8 \text{ l}$$

(g)   Convert 60 miles per hour to meters per second.

Conversion factors needed:

$$1 \text{ mile} = 1.6 \text{ km}$$
$$1 \text{ km} = 10^3 \text{ m}$$
$$1 \text{ hr} = 60 \text{ min}$$
$$1 \text{ min} = 60 \text{ sec}$$

therefore,

$$\frac{60 \text{ mi}}{\text{hr}} \times \frac{1.6 \text{ km}}{\text{mi}} \times \frac{10^3 \text{ m}}{\text{km}} \times \frac{1 \text{ hr}}{60 \text{ min}} \times \frac{1 \text{ min}}{60 \text{ sec}} =$$

$$\frac{(60)(1.6)(10^3)}{(60)(60)} \frac{\text{m}}{\text{sec}} = \frac{9.6 \times 10^4}{3.6 \times 10^3} \frac{\text{m}}{\text{sec}} = 2.7 \times 10^1 \frac{\text{m}}{\text{sec}}$$

In carrying out lengthy conversions such as this one, where two different dimensions are involved, i.e., here both distance and time must be converted, it is easier to completely convert one dimension all the way to the desired units before starting the conversion of the other dimension. Notice that here the units of distance (miles) are converted to the desired units of distance (meters) first, then the units of time are converted from hours to seconds.

(h)   Express one light-year in distance traveled in centimeters. (One light-year is the distance that light travels in one year traveling at a speed of 3×10^{10} centimeters per second.)

Conversion factor needed:

$$1 \text{ yr} = 365 \text{ days}$$
$$1 \text{ day} = 24 \text{ hrs}$$
$$1 \text{ hr} = 60 \text{ min}$$
$$1 \text{ min} = 60 \text{ sec}$$

therefore,

$$3 \times 10^{10} \; \frac{cm}{sec} \; \times \; \frac{60 \text{ sec}}{1 \text{ min}} \; \times \; \frac{60 \text{ min}}{1 \text{ hr}} \; \times \; \frac{24 \text{ hrs}}{1 \text{ day}} \; \times \; \frac{365 \text{ days}}{1 \text{ yr}} =$$

$$(3 \times 10^{10}) \; (3.6 \times 10^{3}) \; (2.4 \times 10^{1}) \; (3.65 \times 10^{2}) \; \frac{cm}{yr} =$$

$$9.4 \times 10^{17} \; \frac{cm}{yr}$$

(i)   Convert furlongs per fortnight to speed in centimeters per second.

Conversion factors needed:

1 furlong	= 220 yards	1 fortnight = 14 days	
1 yard	= 3 feet	1 day	= 24 hours
1 foot	= 12 inches	1 hour	= 60 minutes
1 inch	= 2.54 cm	1 minute	= 60 seconds

therefore,

$$\frac{1 \text{ furlong}}{\text{fortnight}} \; \times \; \frac{220 \text{ yd}}{1 \text{ furlong}} \; \times \; \frac{3 \text{ ft}}{1 \text{ yd}} \; \times \; \frac{12 \text{ in}}{1 \text{ ft}} \; \times \; \frac{2.54 \text{ cm}}{1 \text{ in}} \; \times \; \frac{1 \text{ fortnight}}{14 \text{ days}}$$

$$\times \; \frac{1 \text{ day}}{24 \text{ hr}} \; \times \; \frac{1 \text{ hr}}{60 \text{ min}} \; \times \; \frac{1 \text{ min}}{60 \text{ sec}} \; = \; 1.66 \times 10^{-2} \; \frac{cm}{sec}$$

(j)   Calculate the number of carbon atoms present in 5.0 grams of
       carbon.

       Conversion factors needed:

              12 g carbon   = 1.0 mole of C atoms

              1.0 mole of C atoms   = 6.0 x 10^{23} C atoms

       therefore,

       5.0 g carbon x $\dfrac{1.0 \text{ mole}}{12 \text{ g carbon}}$ x $\dfrac{6.0 \times 10^{23} \text{ C atoms}}{1.0 \text{ mole}}$

              = 2.5 x 10^{23} C atoms

The conversion of units of area and volume require special attention in
some cases. A square that measures 6 inches on each side has an area
of 36 inches, i.e., 36 in^2. Volume in the metric system is derived from
the fundamental unit of length, i.e., a cube which has a length of one
centimeter on its edge has a volume of one centimeter cubed or 1 cm^3.
When units of length are used in making conversions of area and volume,
the entire conversion factor, i.e., the number and the unit must be either
squared or cubed depending upon the conversion.

Examples:

(a)   Express the area of a room which measures 15 ft x 12 ft in square
       inches.

       (15 ft x 12 ft) = 185 ft^2 (area in ft^2)

       since

              1 ft = 12 in,

       185 ft^2 x $\left(\dfrac{12 \text{ in}}{1 \text{ ft}}\right)^2$ =   185 ft^2 x $\dfrac{144 \text{ in}^2}{1 \text{ ft}^2}$ = 26,640 in^2

                                                                    = 2.7 x 10^4 in^2

(b)   Calculate the volume in cubic feet of a rectangular box which has the dimensions of 30 mm x 40 mm x 50 mm.

Conversion factors needed:

$$1 \text{ cm} \quad = 10 \text{ mm}$$
$$30.5 \text{ cm} = 1 \text{ foot}$$

therefore,

$$(30 \text{ mm} \times 40 \text{ mm} \times 50 \text{ mm})\left(\frac{1 \text{ cm}}{10 \text{ mm}}\right)^3 \times \left(\frac{1 \text{ ft}}{30.5 \text{ cm}}\right)^3 \quad =$$

$$(6.0 \times 10^4 \text{ mm}^3)\left(\frac{1 \text{ cm}^3}{10^3 \text{ mm}^3}\right) \times \left(\frac{1 \text{ ft}^3}{28,365 \text{ cm}^3}\right) \quad =$$

$$\frac{6.0 \times 10^4}{2.84 \times 10^7} \text{ ft}^3 = 2.1 \times 10^{-3} \text{ ft}^3$$

PROBLEMS

1.   Perform the following conversions.

   (a)   15 dm to meters

   (b)   5.2 kg to grams

   (c)   2 ml to liters

   (d)   75 cm to millimeters

   (e)   540 mm to kilometers

   (f)   50 kg to centigrams

   (g)   45 dg to kilograms

   (h)   14 g to micrograms

   (i)   75 miles to kilometers

   (j)   2.5 A to feet

   (k)   3 inches to microns

   (l)   15 mμ to centimeters

   (m)   25 lb to picograms

   (n)   12 miles to terameters

   (o)   5 quarts to liters

   (p)   20 gallons to milliliters

   (q)   21 quarts to microliters

   (r)   10 miles/hr to m/sec

    (s)   5 A to nanometers

    (t)   1500 cm^3 to liters

    (u)   2 fortnights to gigaseconds

    (v)   40 g to tons (U.S.)

    (w)   2 metric tons to ounces

    (x)   40 ft^2 to in^2

    (y)   25 cm^3 to mm^3

    (z)   300 mm^3 to quarts

2.   Express the speed of a car traveling 65 miles per hour in centimeters per second.

3.   Calculate the area in cm^2 of a rectangle which measures 15 in x 10 in.

4.   Calculate the volume of a cube which is 6.00 cm on each edge.

5.   Calculate the volume in cm^3 of a box which is 2 in x 5 in x 4 in.

6.   Calculate the area in cm^2 of a square which is 5 A on each side.

7.   Convert 100,000 centimeters per second to fathoms per fortnight. (1 fathom = 1.83 meters)

8.   Convert 250 liters to firkins. (1 firkin = 9 gallons)

9.   A strip of metal has a length of 2.75 cm. Express this length in:

    (a)   mm
    (b)   inches
    (c)   Angstrom
    (d)   yards

10.   The volume of a sodium atom is approximately $27A^3$.   Express
       this volume in:

       (a) in^3
       (b)   ml
       (c) ft^3
       (d)   liters

Unit 4

THE SLIDE RULE

The slide rule is an instrument used extensively in the physical sciences for performing certain arithmetical calculations quickly and accurately. Any calculation in arithmetic except addition and subtraction can be performed using the slide rule. The most common operations are multiplication, division and combinations of these; the square of a number or its square root; the cube of a number or its cube root; and operations involving logarithms.

The slide rule consists of three parts: the fixed part called the body which usually has the A, D, K, and L scales; the slide, or movable part in the center, which ordinarily has the B, C, and CI scales; and the indicator which consists of a window with a vertical hairline. The indicator slides in grooves on the outer edges of the body of the slide rule. The different parts of a slide rule are illustrated in Figure 4-1.

PRINCIPLE OF OPERATION OF THE SLIDE RULE

It is important to understand the principle upon which the operation of the slide rule is based. The process of multiplication of two numbers on the slide rule corresponds to the addition of the logarithms (to be discussed in Unit 5) of these numbers. The division process corresponds to the subtraction of the logarithms of numbers. To understand how this is done, examine the scales on the slide rule labeled C and D. These scales are identical except that the C scale is located on the movable part (the slide) and the D scale is on the stationary part (the body) of the slide rule. On both of these scales, the position of the numbers are determined by the values of their logarithms. For example, the number 3 is located about 50 per cent of the distance from the left index (1) to the right index (10). This is due to the fact that the logarithm of 3, i.e., 0.4771, lies 47.71 per cent of the distance between the logarithm of 1 (0.0000) and the logarithm of 10 (1.0000).

41

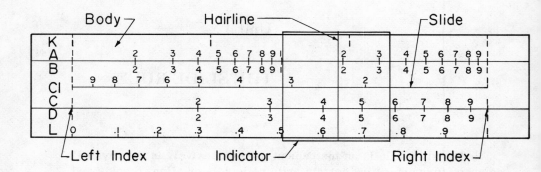

Figure 4-1.   Parts of the Slide Rule.

To illustrate the multiplication process, consider the slide rule setting shown in Figure 4-2.

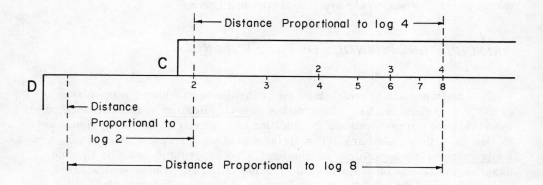

Figure 4-2.   Multiplication: 2 x 4 = 8.

Division: 8 ÷ 4 = 2.

In Figure 4-2, the left index of the C scale is placed directly above
the number 2 on the D scale and the number 4 on the C scale is directly
above 8 on the D scale. By this simple procedure, the multiplication process, 2 x 4 = 8, has been carried out. Notice that actually two <u>distances</u>
have been <u>added</u>. One of these distances is proportional to the logarithm
of 2, i.e., 0.3010, and the other distance is proportional to the logarithm
of 4, i.e., 0.6021. The total distance is proportional to the logarithm of
8, i.e., 0.9031. In other words, two numbers are multiplied using the
slide rule by the addition of their logarithms.

Figure 4-2 can also be used to illustrate division using the slide rule.
Remember that the division process corresponds to the <u>subtraction</u> of the
logarithms of numbers. In Figure 4-2, since the number 4 on the C scale
is directly above 8 on the D scale, the left index of the C scale is necessarily
above 2 on the D scale, i.e., the division process, 8 ÷ 4 = 2, has been performed. Here, the distance which is proportional to the logarithm of 4,
i.e., 0.6021, has been <u>subtracted</u> from the distance proportional to the
logarithm of 8, i.e., 0.9031, resulting in the distance which is proportional
to the logarithm of the quotient 2, i.e., 0.3010.

READING THE SCALES ON THE SLIDE RULE

The most important aspect of learning to use the slide rule properly
is the accurate reading of the scales. When this has been accomplished
for one of the scales, only a little extra time and effort are needed to
master the reading of all the scales. To correctly read the scales on the
slide rule, it is necessary to be thoroughly familiar with the major and
minor divisions of each whole number. The whole number divisions on the
C and D scales are shown in Figure 4-3.

Figure 4-3.   Whole number divisions, C and D scales.

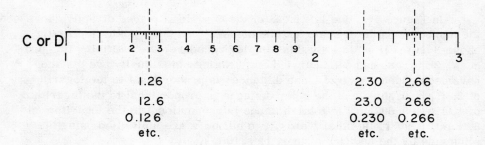

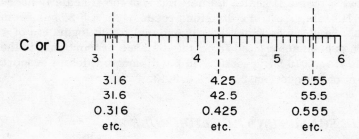

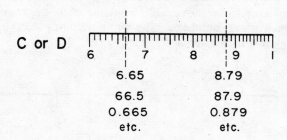

Figure 4-4.   Location of Numbers on the C or D Scales.

Notice that the whole number divisions 1, 2, 3, 4, 5, 6, 7, 8, 9, and 10 appear at progressively smaller intervals moving from left to right. This compression of the space between the numbers is a result of the logarithmic nature of the scales. The space between each whole number is first divided into tenths, and further division of the tenths between each whole number varies with the size of the space between the numbers. The space between 1 and 2, since it is larger, permits, as illustrated in Figure 4-4, each tenth to be further divided into tenths (marked with small numerals 1 through 9), for a total of 100 divisions between the whole numbers 1 and 2. The space between the whole numbers 2 and 4 is divided into tenths and then further divided into fifths between the tenths making a total of 50 divisions each between 2 and 3 and between 3 and 4. . Between the whole number 4 to 10 each number is divided into tenths, and each tenth is then divided into halves making a total of 20 divisions between each whole number from 4 to 10. Notice that in view of the above considerations, both the setting and reading of numbers becomes less precise moving from left to right on the slide rule, i.e., between 1 and 2, the first three digits can be read directly and the fourth estimated, whereas between 4 and 10 the third digit must be estimated by the spacing between the tenths if it is any number other than 5.

In locating a number on the slide rule, it is necessary to consider only the sequence of the digits, not the magnitude of the number. Any number beginning with 1, regardless of magnitude, is located between the large 1 and the large 2 on the C and D scales. Any number beginning with 2 is located between the large 2 and the large 3, etc. The location of several numbers on various parts of the slide rule is illustrated in Figure 4-4. The correct positions of these numbers are indicated by dotted vertical lines.

USING THE SLIDE RULE

Multiplication

Multiplication and division are the most important operations carried out using the slide rule because they are the most frequently performed. The C and D scales are ordinarily used for both multiplication and division. The simplest operation is that of multiplying one number, the multiplicand, by another number, the multiplier. The process is generally carried out in the following steps:

(1)   The first number, or multiplicand, is located on the D scale by adjusting the slide so that one index of the C scale is directly over the number.

(2)   The indicator is then placed so that the hairline coincides with the second number, or multiplier, on the C scale.

(3)   The product is then read on the D scale under the hairline.

Several examples will illustrate the above steps.

Examples

Perform the following operations :

(a)   2.50 x 3.10 =

   (1)   Adjust the slide so that the left index of the C scale is directly over 2.50 on the D scale.

   (2)   Slide the hairline to 3.10 on the C scale.

   (3)   Read the answer, 7.75, under the hairline on the D scale. (See Figure 4-5.)

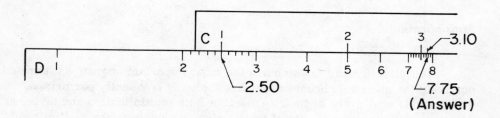

Figure 4-5.   Multiplication: 2.50 x 3.10 = 7.75

(b)   9.5 x 6.0 =

   (1)   Adjust the slide so that the right index of the C scale is
          directly over 9.5 on the D scale.

   (2)   Slide the hairline to 6.0 on the C scale.
          (Note that when the left index is used in step (1), 6.0 is off
          scale to the right.)

   (3)   Read the answer, 57, under the hairline on the D scale.
          (See Figure 4-6.)

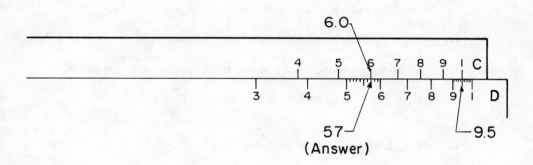

Figure 4-6.   Multiplication: 9.5 x 6.0 = 57

(c)   2.70 x 4.20 x 3.10 =

   (1)   Adjust the slide so that the right index of the C scale is
          directly over 2.70 on the D scale.

   (2)   Slide the hairline to 4.20 on the C scale. (Notice that the
          product of these two numbers 11.34, could be read on the D
          scale at this point. However, since there is an additional
          step in this problem, it is not necessary to record the product
          of the first two numbers.)

(3)   Adjust the slide so that the left index of the C scale coincides
       with the hairline, i.e., above 11.34 on the D scale.

(4)   Slide the hairline to 3.10 on the C scale.

(5)   Read the answer, 35.2, under the hairline on the D scale.

Notice that this process could be continued indefinitely depending upon the
length of the problem, i.e., the product, 35.2, can now be multiplied by
any number desired. To multiply 35.2 by 2.5, continue the operation as
follows:

(6)   Adjust the slide so that the left index of the C scale coincides
       with the hairline, i.e., above 35.2 on the D scale.

(7)   Slide the hairline to 2.50 on the C scale.

(8)   Read the answer, 88.0, under the hairline on the D scale.

       (i.e., 2.70 x 4.20 x 3.10 x 2.50 = 88.0)

(d)   (275,000) (1,480) (0.000,00120) =

Operations such as this one are most easily performed by the use of
exponential notation and a "common sense" approach to the location of the
decimal point, i.e., rewrite the problem using exponential notation as
follows:

$$(2.75 \times 10^5) \ (1.48 \times 10^3) \ (1.20 \times 10^{-6}) \ =$$

The location of the decimal in the final answer can be arrived at by mentally
carrying out an "order of magnitude" or "ballpark" type calculation, i.e.,
(3) (1.5) (1) x 10^5 x 10^3 x 10^{-6} = 4.5 x 10^2, i.e., the decimal point is located
so that the coefficient in the final answer is closer to 4.5 than it is to 0.45
or 45.

Now, using the slide rule, the exact answer can be determined as illustrated above, i.e.,

$$(2.75) (1.48) (1.20) \times 10^5 \times 10^3 \times 10^{-6} =$$

(1)   Adjust the slide so that the <u>left</u> index of the C scale is directly over 2.75 on the D scale.

(2)   Slide the hairline to 1.48 on the C scale.

(3)   Adjust the slide so that the <u>left index</u> of the C scale coincides with the <u>hairline.</u> (That is, above 4.07 on the D scale.)

(4)   Slide the hairline to 1.2 on the C scale.

(5)   Read the answer, 4.89, under the hairline on the D scale.

Therefore, the final answer is: 4.89×10^2. (Not 0.489×10^2 or 48.9×10^2 in view of the "ballpark" calculation carried our initially.)

Division

In the division process, a number, the <u>dividend</u> is divided <u>by</u> another another number, the <u>divisor.</u> The answer or result is called the <u>quotient.</u> The C and D scales of the slide rule are ordinarily used to perform operations involving the division of numbers also. Remember, however, that the division process using the slide rule corresponds to the <u>subtraction</u> of the <u>logarithms</u> of numbers, i.e., the subtraction of <u>distances</u> which are proportional to the logarithms of numbers. The division process is generally carried out in the following steps:

(1)   The <u>hairline</u> of the indicator is placed so that it coincides with the dividend, i.e., the number to be divided, on the D scale.

(2)   The slide is adjusted so that the divisor, the number which divides into the dividend, is located on the C scale under the hairline, i.e., directly over the dividend.

(3)   The answer, or quotient, is read on the D scale below the on-scale index of the C scale.

Several examples will illustrate the above steps. (Note that the operations for division are the reverse of those for multiplication.)

Examples

Perform the following operations:

(a) $\dfrac{35.6}{4.90}$ =   (mental arithmetic indicates that the answer is $\dfrac{35}{5}$, or approximately 7)

    (1)   Place the hairline of the indicator over 35.6 on the D scale.

    (2)   Adjust the slide so that 4.90 on the C scale is directly under the hairline, i.e., 4.90 on the C scale is directly over 35.6 on the D scale.

    (3)   Read the answer, 7.27, on the D scale below the right index of the C scale.

    (See Figure 4-7.)

(b) $\dfrac{386}{24.9}$ =   (approximate answer is $\dfrac{400}{25}$ or 16)

    (1)   Place the hairline of the indicator over 386 on the D scale.

    (2)   Adjust the slide so that 24.9 on the C scale is directly under the hairline.

    (3)   Read the answer, 15.5, on the D scale below the left index of the C scale. (See Figure 4-8.)

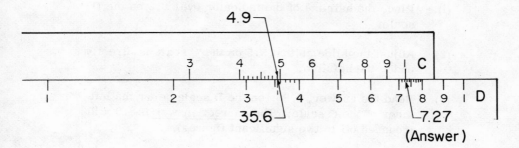

Figure 4-7.   Division: 35.6 ÷ 4.90 = 7.27

(c) $\dfrac{65,000}{0.000,035}$ =

rewritten in exponential notation, the problem is:

$\dfrac{6.5 \times 10^{4}}{3.5 \times 10^{-5}}$ =   (approximate answer is 2×10^{9})

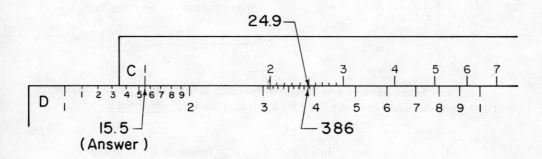

Figure 4-8.   Division: 386 ÷ 24.9 = 15.5

(1)   Place the hairline of the indicator over 6.5 on the D
      scale.

(2)   Adjust the slide so that 3.5 on the C scale is directly
      under the hairline.

(3)   Read the answer, 1.86, on the D scale below the left
      index of the C scale. The <u>correct answer</u> is 1.9×10^9
      (rounded off to two significant figures).

Combination of Multiplication and Division

Frequently, it is necessary to perform calculations involving both
multiplication and division. Problems of this type can be rapidly solved
in one continuous operation using the slide rule. The method described
below is the easiest for beginners to learn. As proficiency and confidence
are developed by using the slide rule, more sophisticated (and less cumber-
some) methods can be mastered.

Examples

(a) $\dfrac{3.9 \times 5.1 \times 2.3}{1.9 \times 4.2 \times 3.4}$ =   (approximate answer =

$$\frac{4 \times 5 \times 2}{2 \times 4 \times 3} = \frac{40}{24} = 2)$$

The exact answer can be found by first determining the numerator
product and then dividing by 1.9, 4.2, and 3.4 <u>in sequence</u>.

(1)   Adjust the slide so that the <u>right</u> index of the C scale
      is directly over 3.9 on the D scale.

(2)   Slide the hairline to 5.1 on the C scale.

(3)   Adjust the slide so that the <u>left</u> index of the C scale
      coincides with the <u>hairline</u>.

(4)   Slide the hairline to 2.3 on the C scale. (The numerator product is 45.75. Now the division process can be carried out.)

(5)   Adjust the slide so that 1.9 on the C scale is directly under the hairline.

(6)   Slide the hairline to the _left_ index of the C scale.

(7)   Adjust the slide so that 4.2 on the C scale is directly under the hairline.

(8)   Slide the hairline to the right index of the C scale.

(9)   Adjust the slide so that 3.4 on the C scale is directly under the hairline.

(10) Read the answer, 1.685 (not in significant figures), on the D scale below the left index of the C scale.

(b) $\dfrac{525 \times 1700 \times 0.005}{1250 \times 0.0002 \times 75}$ =

rewritten in exponential notation, the problem is:

$$\frac{(5.25 \times 10^{2})\,(1.7 \times 10^{3})\,(5 \times 10^{-3})}{(1.25 \times 10^{3})\,(2 \times 10^{-4})\,(7.5 \times 10^{1})} =$$

The approximate answer is:

$$\frac{(5)\,(2)\,(5) \times 10^{2+3-3}}{(1)\,(2)\,(8) \times 10^{3-4+1}} = \frac{50}{16} \times 10^{2} = 3 \times 10^{2}$$

Following the procedure described above, the slide rule digit sequence 238. Therefore, the exact answer is 2.38×10^{2} (not in significant figures).

Squares and Square Roots

In squaring a number or extracting the square root of a number, the A and D scales of the slide rule are used. Notice that the A scale is a double scale which has the same numbers as the C or D scale on each half. The <u>left</u> half of the A scale is used when extracting the square root of a number with an <u>odd number of digits</u> to the left of the decimal, while the <u>right</u> half of the A scale is used extracting the square root of a number with an <u>even number of digits</u> to the left of the decimal. When extracting the square roots of very large and very small numbers, the number should be first rewritten using exponential notation. Recall from Unit 1 that in order to extract the square root of a number written in exponential form, that the <u>power of ten must be divisible by two</u> to give a whole number, and that the <u>coefficient</u> should never be less than one. The reason for following this convention will be explained in the examples below.

<u>Procedure:</u>   To extract the square root of a number, set the hairline of the indicator over the number on the A scale and read the square root under the hairline on the D scale.

Examples

(a) $(4.0)^{1/2} = 2.0$

The hairline is set over 4 on the <u>left</u> half of the A scale (i.e., odd number of digits to left of decimal) and 2.0 is read under the hairline on the D scale. (See Figure 4-9.)

(b) $(6.25)^{1/2} = 2.50$

The hairline is set over 6.25 on the <u>left</u> half of the A scale and the square root, 2.50, is read under the hairline on the D scale. (See Figure 4-9.)

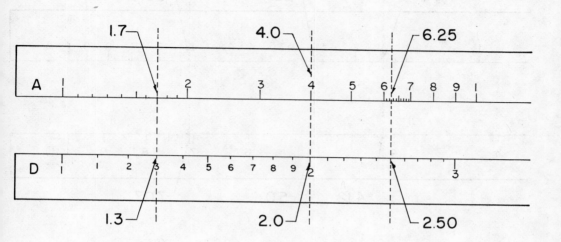

Figure 4-9.   Square roots of numbers: $(4.0)^{1/2} = 2.0$;
$(6.25)^{1/2} = 2.50$; $(1.7 \times 10^4)^{1/2} = 1.3 \times 10^2$.

(c) $(25)^{1/2} = 5.0$

The hairline is set over 25 on the <u>right</u> half of the A scale
(i.e., even number of digits to the left of decimal) and 5.0 is
read under the hairline on the D scale. (See Figure 4-10.)

(d) $(50.0)^{1/2} = 7.07$

The hairline is set over 50.0 on the <u>right</u> half of the A scale
and the square root, 7.07, is read under the hairline on the
D scale. (See Figure 4-10.)

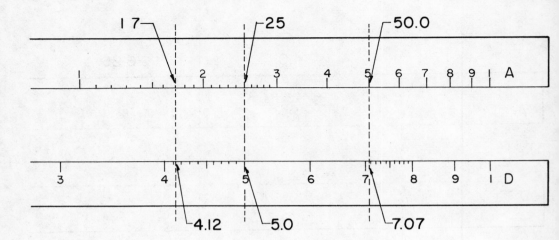

Figure 4-10.   Square roots of numbers: $(25)^{1/2} = 5.0$;
$(50.0)^{1/2} = 7.07$; $(17 \times 10^4)^{1/2} = 4.12 \times 10^2$.

(e) $(17,000)^{1/2} = ?$

rewritten in exponential form <u>so that the power of ten is</u>
<u>divisible by</u> 2, the problem is:

$(1.7 \times 10^4)^{1/2} = ? \times 10^2$

To extract the square root of the coefficient, the hairline is set
over 1.7 on the <u>left</u> half of the A scale and the square root, 1.3,
is read under the hairline on the D scale. That is,

$(1.7 \times 10^4)^{1/2} = 1.3 \times 10^2$

(See Figure 4-9.)

(f) $(170,000)^{1/2} = ?$

rewritten in exponential form so that the <u>power of ten is divisible by 2,</u> and the coefficient is greater than one,

$(17 \times 10^4)^{1/2} = ? \times 10^2$

To extract the square root of the coefficient, the hairline is set over 17 on the <u>right</u> half of the A scale and the square root, 4.12, is read under the hairline on the D scale, i.e.,

$(17 \times 10^4)^{1/2} = 4.12 \times 10^2$

(See Figure 4-10.)

(g) $(0.000,000,023)^{1/2} = ?$

$(2.3 \times 10^{-8})^{1/2} = ? \times 10^{-4}$

To extract the square root of the coefficient, set the hairline over 2.3 on the <u>left</u> half of the A scale and read the root under the hairline on the D scale. That is,

$(2.3 \times 10^{-8})^{1/2} = 1.515 \times 10^{-4}$

(Not in significant figures.)

Squaring a number using the slide rule is the reverse of the pro-cedure for extracting square roots, i.e., the hairline of the indicator is set over the number on the D scale and its square is read on the A scale. Again, exponential notation should be used when working with very large and very small numbers.

Examples

(a) $(8.5)^2 = 72.25$ (Not in significant figures.)

The hairline is set over 8.5 on the D scale, and the answer,
72.25, is read under the hairline on the A scale. Notice that
the answer is on the <u>right</u> half of the A scale, i.e., there are
<u>two</u> digits to the left of the decimal in the answer.

(b) $(27,000)^2 = ?$

$(2.7 \times 10^4)^2 = ? \times 10^8$

To square the coefficient, the hairline is set over 2.7 on the
D scale, and its square is read under the hairline on the A
scale. (Left half of the A scale, therefore, one digit to the
left of the decimal in the coefficient.) That is,

$(2.7 \times 10^4)^2 = 7.3 \times 10^8$

(c) $(0.00\ 52)^2 = ?$

$(5.2 \times 10^{-4})^2 = ? \times 10^{-8}$

To square the coefficient, set the hairline over 5.2 on the
D scale and read the answer 27 under the hairline on the <u>right</u>
half of the A scale. (Right half of A scale, therefore, two
digits to the left of decimal in the coefficient.) That is,

$(5.2 \times 10^{-4})^2 = 27 \times 10^{-8}$

$$= 2.7 \times 10^{-7}$$

Cubes and Cube Roots

In cubing a number or extracting the cube root of a number, the K
and D scales of the slide rule are used. The K scale is a triple scale

which has the same numbers as the C or D scale on each third. The <u>left</u>
third of the K scale is used when extracting the cube root of a number with
<u>one digit</u> to the left of the decimal, the <u>middle</u> third is used when extracting
the cube root of a number with <u>two digits</u> to the left of the decimal, and the
<u>right</u> third is used when extracting the cube root of a number with <u>three</u>
<u>digits</u> to the left of the decimal. When extracting the cube roots of very
large and very small numbers, the numbers should be written in exponential
form so that the <u>power of ten is divisible by three</u> to give a whole number,
and the coefficient is a number between one and one thousand (i.e., one,
two, or three digits to the left of the decimal). Several examples will
illustrate the above points.

Examples

(a) $(8.50)^{1/3} = 2.04$

The hairline is set over 8.5 on the <u>left</u> third of the K scale (i.e.,
<u>one</u> digit to the left of the decimal), and the cube root, 2.04, is
read under the hairline on the D scale.

(b) $(27.5)^{1/3} = 3.02$

The hairline is set over 27.5 on the <u>middle</u> third of the K scale
(i.e., <u>two</u> digits to the left of the decimal), and the cube root,
3.02, is read under the hairline on the D scale.

(c) $(125)^{1/3} = 5$

The hairline is set over 125 on the <u>right</u> third of the K scale
(i.e., <u>three</u> digits to the left of the decimal), and the answer,
5, is read under the hairline on the D scale.

(d) $(18,600,000)^{1/3} = ?$

rewritten in exponential form so that the <u>power of ten is divisible</u>
<u>by three</u> and so the coefficient has one, two, or three digits to the
left of the decimal,

$$(18.6 \times 10^6)^{1/3} = ? \times 10^2$$

To extract the cube root of the coefficient, the hairline is set over 18.6 on the <u>middle</u> third of the K scale and the cube root, 2.65, is read under the hairline on the D scale, i.e.,

$$(18.6 \times 10^6) = 2.65 \times 10^2$$

(e) $(0.003)^{1/3} = ?$

$$(3 \times 10^{-3})^{1/3} = ?$$

To extract the cube root of the coefficient, the hairline is set over 3 on the <u>left</u> third of the K scale and the answer, 1.44, is read under the hairline on the D scale, i.e.,

$$(3 \times 10^{-3})^{1/3} = 1.44 \times 10^{-1} \quad \text{(not in significant figures)}$$

Cubing a number using the slide rule is the reverse of the procedure for extracting cube roots, i.e., the hairline of the indicator is set over the number on the D scale and its cube is read on the A scale. The most convenient method for handling very large and very small numbers is, as usual, exponential notation.

Examples

(a) $(2.10)^3 = 9.26$

The hairline is set over 2.10 on the D scale, and its cube, 9.26, is read under the hairline on the <u>left</u> third of the K scale. (Since the answer is on the left third of the K scale, there is <u>one</u> digit to the left of the decimal in the answer.)

(b) $(0.0038)^3 = ?$

$$(3.8 \times 10^{-3})^3 = ? \times 10^{-9}$$

To cube the coefficient, the hairline is set over 3.8 on the D scale. The answer, 54.9, is read under the hairline on the middle third of the K scale. (Therefore, there are two digits to the left of the decimal in the coefficient.) That is,

$$(3.8 \times 10^{-3})^3 = 54.9 \times 10^{-9}$$

$$= 5.49 \times 10^{-8} \text{ (not in significant figures)}$$

(c) $(51,000,000)^3 = ?$

$(5.1 \times 10^7)^3 = ? \times 10^{21}$

To cube the coefficient, set the hairline over 5.1 on the D scale and read the answer, 133, on the right third of the K scale, i.e., there are three digits to the left of the decimal in the coefficient.

Therefore,

$$(5.1 \times 10^7)^3 = 133 \times 10^{21}$$

$$= 1.3 \times 10^{23}$$

PROBLEMS

1.   Perform the following multiplications and divisions using the
     slide rule. First, <u>approximate</u> the answer in order to locate the
     decimal place. Use exponential notation when appropriate.

     (a)   (15) (28)

     (b)   (3.5) (4.2)

     (c)   (16.8) (2.4)

     (d)   (1.9) (8.4) (1.3)

     (e)   (2.9) (4.6) (1.4) (8.6)

     (f) $(1.4 \times 10^5) (2.6 \times 10^3)$

     (g) $(3.3 \times 10^{-4}) (1.8 \times 10^{-5})$

     (h) $(6.5 \times 10^2) (2.1 \times 10^6)$

     (i) $(1.7 \times 10^{-2}) (4.9 \times 10^5) (3.2 \times 10^{-8})$

     (j) $\dfrac{157}{24}$

     (k) $\dfrac{3,250}{65}$

     (l) $\dfrac{49,500}{175}$

     (m) $\dfrac{(2.8 \times 10^4)}{(1.7 \times 10^2)}$

     (n) $\dfrac{(1.5 \times 10^{-3})}{(7.5 \times 10^{-2})}$

(o) $\dfrac{(7.2 \times 10^{25})}{(4.9 \times 10^{-20})}$

(p) $(0.000,50)\ (275,000)$

(q) $(1,800,000)\ (0.000,000,25)$

(r) $\dfrac{0.000,027}{185,000}$

(s) $\dfrac{23,500}{0.000,89}$

2.   Perform the following operations using the slide rule.

(a) $\dfrac{(275)\ (14.8)\ (6.2)}{(1.25)\ (14.3)}$

(b) $\dfrac{(1.75)\ (2.34)\ (82)}{(14)\ (1.23)\ (0.55)}$

(c) $\dfrac{(1.75 \times 10^{3})\ (2.8 \times 10^{-5})}{(7.2 \times 10^{-6})}$

(d) $\dfrac{(4.9 \times 10^{7})}{(2.6 \times 10^{4})\ (1.3 \times 10^{14})}$

(e) $\dfrac{(7.7 \times 10^{5})\ (9.4 \times 10^{6})}{(6.3 \times 10^{4})\ (1.5 \times 10^{-2})}$

(f) $\dfrac{(285,000)\ (1,760)}{(14,000)\ (0.0035)}$

(g) $\dfrac{(0.000,47)\ (0.000,000,29)}{(950,000)\ (275)\ (1920)}$

3.   Find the indicated powers or roots of each of the following.

    (a) $(2.35)^2$

    (b) $(3.1)^2$

    (c) $(5.2)^2$

    (d) $(1.6)^3$

    (e) $(3.5)^3$

    (f) $(45)^3$

    (g) $(0.000,0048)^2$

    (h) $(0.000,025)^3$

    (i) $(185,000)^2$

    (j) $(19,000,000)^2$

    (k) $(85,000)^3$

    (l) $(1,230,000,000)^3$

    (m) $(16.5)^{1/2}$

    (n) $(2.9)^{1/2}$

    (o) $(183)^{1/2}$

    (p) $(0.000,0048)^{1/2}$

    (q) $(93,000,000)^{1/2}$

    (r) $(1.75)^{1/3}$

    (s) $(17.5)^{1/3}$

    (t) $(175)^{1/3}$

    (u) $(286,000)^{1/3}$

(v) $(0,000,000,0029)^{1/3}$

4.   Perform the indicated mathematical operations using the slide
      rule.

(a) $(8.5 \times 10^4)^{1/2}(6.2 \times 10^{-2})^2$

(b) $(2.8 \times 10^{-3})^{1/2}(1.7 \times 10^4)^3$

(c) $\dfrac{(9.4 \times 10^{-6})^{1/2}}{(4.3 \times 10^4)^{1/2}}$

(d) $\dfrac{(1.94 \times 10^{-6})^{1/3}}{(2.7 \times 10^{-3})^{1/3}}$

(e) $\dfrac{(7.5 \times 10^3)\ (3.8 \times 10^5)^{1/2}}{(9.1 \times 10^{-2})}$

(f) $\dfrac{(5.5 \times 10^{20})^{1/3}(1.8 \times 10^{-3})}{(7.25 \times 10^3)^{1/3}}$

(g) $(0.000,000,125)^{1/3}(850,000)^2$

(h) $\dfrac{(1,750,000)\ (0.000,000,000,28)^3(145,000)^{1/2}}{(0.000,000,000,45)^{1/3}(8,750)(0.000,0095)^2}$

(i) $\dfrac{(15,000)^2\ (0.000,075)\ (0.000,23)^{1/2}}{(14,000,000)\ (0.000,0045)}$

(j) $\dfrac{(3,750)^2\ (0.000,000,065)^{1/3}\ (0.000,025)}{(14,250)\ (72,000)^{1/2}\ (0.000,000,48)}$

(k) $\dfrac{(1,450)^3\ (0.000,000,72)^{1/3}\ (0.000,075)^2}{(180,000)\ (23,500)\ (64,000)^{1/2}}$

(l) $\left[\dfrac{(1.8 \times 10^{6})(2.9 \times 10^{-4})}{(7.5 \times 10^{-2})(8.2 \times 10^{5})}\right]^{2}$

(m) $\left[\dfrac{(8.3 \times 10^{-5})(7.9 \times 10^{-6})}{(2.4 \times 10^{4})}\right]^{1/2}$

(n) $\left[\dfrac{(83,000)(950,000)}{(0.000,045)}\right]^{3}$

(o) $\left[\dfrac{(9.55 \times 10^{6})(1.6 \times 10^{-3})}{(4.7 \times 10^{-2})(1.72 \times 10^{4})}\right]^{1/3}$

Unit 5

LOGARITHMS

COMMON LOGARITHMS

The common logarithm of a number is the power to which 10 (the base) must be raised to equal that number, i.e., a common logarithm is an exponent or power of 10. For example, $100 = 10^2$, therefore, the logarithm of 100 to the base 10 is 2. Similarly, 4 is the common logarithm of 10,000 because $10^4 = 10,000$. Following this line of reasoning, since $10^0 = 1$, then the logarithm of $1 = 0$ and the logarithm of $10 = 1$ because $10^1 = 10$. From Unit 1, recall that 0.001, when expressed as a power of 10, is written as 10^{-3}. Therefore the logarithm to the base 10 of 0.001 is -3. Further examples of some numbers and their common logarithms are given in Table 5-1.

Since $10^0 = 1$ and $10^1 = 10$, it follows that when 10 is raised to a power between zero and one, then a number between 1 and 10 should result. This is, indeed, the case, i.e., $10^{0.6021} = 4$ or stated differently, the logarithm to the base 10 of $4 = 0.6021$. The common logarithm of the numbers between one and ten are given in Table 5-2.

Most logarithms encountered in general chemistry are more involved than the above examples. The logarithms of numbers actually consist of two parts, the characteristic (or whole number part) and the mantissa (the decimal part).

THE CHARACTERISTIC

The characteristic of the logarithm of a number greater than one is one less than the number of digits to the left of the decimal point. Notice that the characteristic is the same as the power of ten when the

TABLE 5-1

Number	Exponential Form	Common Logarithm (base 10)
10,000	10^4	4
1,000	10^3	3
100	10^2	2
10	10^1	1
1	10^0	0
0.1	10^{-1}	-1
0.01	10^{-2}	-2
0.001	10^{-3}	-3
0.0001	10^{-4}	-4

TABLE 5-2

Number	Exponential Form	Common Logarithm (base 10)
1	$10^{0.000}$	0.000
2	$10^{0.301}$	0.301
3	$10^{0.477}$	0.477
4	$10^{0.602}$	0.602
5	$10^{0.699}$	0.699
6	$10^{0.778}$	0.778
7	$10^{0.845}$	0.845
8	$10^{0.903}$	0.903
9	$10^{0.954}$	0.954
10	$10^{1.000}$	1.000

number is written in standard exponential notation, i.e.,

Number	Exponential Notation	Characteristic
7,500	7.5×10^3	3
50	5×10^1	1
3	3×10^0	0
1,850,000	1.85×10^6	6

The characteristic of the logarithm of a number less than one is negative and is one more than the number of zeros to the right of the decimal. Again, the characteristic is the same as the power of ten when the number is written in standard exponential notation, i.e.,

Number	Exponential Notation	Characteristic
0.002	2×10^{-3}	-3
0.4	4×10^{-1}	-1
0.00027	2.7×10^{-4}	-4
0.00000802	8.02×10^{-6}	-6

Note that the characteristic of the logarithm of a number indicates only the magnitude of the number. As discussed below, the mantissa of the logarithm of a number is determined only by the sequence of digits in the number and is independent of the position of the decimal point.

THE MANTISSA

The mantissa or decimal part of the logarithm of a number is always positive and found in logarithm tables such as the one given in the Appendix. The mantissa of the logarithm of any three-digit number can be determined directly to four decimal places using this table of logarithms. If the mantissa of a four-digit number is desired, it is necessary to use the "proportional parts" section of the log table.

To find the mantissa of a number such as 3.85, locate 38 in vertical column N of the log table, then move across to column 5. The mantissa given is 5855 at the intersection. Since the characteristic of 3.85 is zero, therefore,

$$\log 3.85 = 0.5855$$

Notice that any numbers with this same sequence of digits have the same mantissa, only the <u>characteristic</u> is different, i.e.,

$$\log 38.5 = 1.5855$$
$$\log 385 = 2.5855$$
$$\log 3.85 \times 10^6 = 6.5855$$

When determining the mantissa of a four-digit number such as 4527, the proportional parts section of the log table must also be used. The mantissa of 452 is 6551. Since 4527 is "0.7 of the way" between 452 and 453, its mantissa should also be about 0.7 of the way between 6551 and 6561. Therefore, looking under 7 in the proportional parts section of the log table, move down until the line in which 452 appears is reached. The number at this intersection, i.e., 7, must be added to the mantissa of 452 to obtain the mantissa of 4527, as :

$$\text{mantissa } 4527 \ = \ 6551 + 7$$
$$= \ 6558$$

Since the characteristic of 4527 is 3, therefore,

$$\log 4527 \ = \ 3.6558$$

Several examples will illustrate the process of finding the logarithm of a number.

Example

(a)   Find the log of 875

Numbers less than one or greater than ten can be handled
most easily by using exponential notation, since the power of ten
is the characteristic as discussed previously.

since, $875 = 8.75 \times 10^2$

therefore $\log (8.75 \times 10^2) = 0.9420 + 2$

$$= 2.9420$$

(b)   Find the log of 278,000.

therefore $\log (2.78 \times 10^5) = 0.4440 + 5$

$$= 5.4440$$

(c)   Find the log of 0.000172.

$$0.000172 = 1.72 \times 10^{-4}$$

$$\log (1.72 \times 10^{-4}) = 0.2355 - 4$$

$$= \bar{4}.2355 \text{ or } 6.2355 - 10$$

Notice that the logarithm of a number between zero and one has a
negative characteristic. To indicate this, a bar is placed over the 4 only
or alternatively the form 6-10 can be used. It is important to realize
that when written in either form that only the characteristic is negative,
i.e., the mantissa is still positive. However, when expressed in the
following manner,

$$\log (1.72 \times 10^{-4}) = 0.2355 - 4$$

$$= -3.7645$$

the entire logarithm is negative. For some types of chemical calculations,
the latter form is more convenient to use. Calculations of this nature
will be illustrated later in this unit.

ANTILOGARITHMS

An antilogarithm is the number which corresponds to a logarithm. Thus, the antilog of 3 means the number which has a logarithm of 3 or 10^3. To determine the number which is represented by a given logarithm, a procedure which is the reverse of the one for finding the logarithm of a number is used.

Examples

(a)   Find the antilog of 2.3284.

The mantissa, 0.3284, is located in the body of the table at the intersection of the horizontal row marked 21 and the vertical column labeled 3.   Therefore, the digit sequence in the number is 213.   Since the characteristic is 2, there are 3 digits to the left of the decimal point in the number or,

$$\log 213 = 2.3284$$

(i.e., 213 is the number which has a logarithm of 2.3284).

There is an alternate procedure for determining antilogs which is especially useful when the numbers are very large or very small. Since common logarithms are actually exponents or powers of ten, this same problem could have been solved in the following manner:

$$\begin{aligned} \text{antilog } 2.3284 &= 10^{2.3284} \\ &= 10^{0.3284} \times 10^2 \\ &= 2.13 \times 10^2 \end{aligned}$$

This method has the added advantage that the answer always appears in standard exponential notation.

(b)   Find the antilog of 14.6821.

$$\text{antilog } 14.6821 = 10^{14.6821}$$

$$= 10^{0.6821} \times 10^{14}$$
$$= 4.81 \times 10^{14}$$

(c)   Find the antilog of -3.7520.

Remember that written in this form, the entire logarithm is negative. Since a table of logarithms contains only positive mantissae, it is necessary to express the quantity so that the mantissa will be positive, i.e.,

$$\text{antilog}\ (-3.7520) = 10^{-3.7520}$$
$$= 10^{0.2480} \times 10^{-4}$$
$$= 1.77 \times 10^{-4}$$

note that since exponents are added algebraically when exponential numbers are multiplied,

$$10^{0.2480} \times 10^{-4} = 10^{-3.7520}$$

OPERATIONS INVOLVING COMMON LOGARITHMS

Multiplication

Since common logarithms are actually exponents of ten, calculations involving the use of logarithms follow the law of exponents. Therefore, the addition of the logarithms of two or more numbers corresponds to the multiplication of these numbers. This principle was mentioned briefly in Unit 4 in relation to the construction of the slide rule. A more formal statement of this principle is: The logarithm of the product of two or more numbers is the sum of the logarithms of these numbers, i.e.,

$$\log(ab) = \log a + \log b$$

Hence, to multiply two or more numbers, add the logarithms of the numbers and determine the antilogarithm of the sum of the logarithms.

Examples

(a)   43.3 x 0.102 x 5.30 = 23.4

log (43.3 x 0.102 x 5.30) = $\log(4.33 \times 10^1)$ +

$\log(1.02 \times 10^{-1})$ + $\log(5.30 \times 10^0)$

$$= (0.6365 + 1) + (0.0086 - 1)$$
$$+ (0.7243 + 0)$$

$$= (0.6365 + 0.0086 + 0.7243)$$

$$= 1.3694$$

antilog 1.3694 = 23.408

(b) (4.170×10^4) (2.30×10^{-3}) = 95.9

log (4.170×10^4) (2.30×10^{-3}) = log (4.170×10^4) +

log (2.30×10^{-3})

$$= (0.6201 + 4) + (0.3617 - 3)$$
$$= 0.9818 + 1$$
$$= 1.9818$$

antilog 1.9818 = 95.9

Division

For the division of numbers using logarithms, the rule is: the logarithm of the quotient of two numbers is equal to the logarithm of the numerator minus the logarithm of the denominator, i.e.,

$$\log \frac{a}{b} = \log\ a\ -\ \log\ b$$

Therefore, to divide numbers, subtract the logarithm of the denominator from the logarithm of the numerator, and determine the antilogarithm of the difference.

Examples

(a) $\dfrac{62.50}{2.180}$ = 28.67

$$\log \frac{62.50}{2.180} = \log 62.50\ -\log 2.180$$

$$= 1.7959 - 0.3385$$
$$= 1.4574$$

antilog 1.4574 = 28.67

(b) $\dfrac{16.4 \times 3.52}{7.38}$ = 7.82

$$\log \frac{16.4 \times 3.52}{7.38} = \log 16.4 + \log 3.52 - \log 7.38$$
$$= 1.2148 + 0.5465 - 0.8681$$
$$= 0.8932$$

antilog 0.8932 = 7.82

Raising Numbers to Powers

To raise a number to a power using logarithms, multiply the logarithm of the number by the power and determine the antilogarithm of the product. The mathematical rule involved here is:

$$\log (a)^n = n \log a$$

Examples

(a) $(5.250)^3 = 144.7$

$$\log (5.250)^3 = 3 \log (5.250)$$
$$= 3(0.7202)$$
$$= 2.1606$$

antilog $2.1606 = 144.7$

(b) $(1.50 \times 10^2)^4 = 5.06 \times 10^8$

$$\log (1.50 \times 10^2)^4 = 4 \log (1.50 \times 10^2)$$
$$= 4(2.1761)$$
$$= 8.7044$$

antilog $8.7044 = 10^{8.7044}$
$$= 10^{0.7044} \times 10^8$$
$$= 5.06 \times 10^8$$

Extracting Roots of Numbers

To find any root of a number using logarithms, divide the logarithm of the number by the root and determine the antilogarithm of the quotient. Here, the mathematical relationship involved is:

$$\log(a)^{1/n} = \frac{1}{n} \log a$$

Examples

(a) $(159)^{1/4}$ = 3.55

$\log (159)^{1/4}$ = 1/4 log 159

                    = 1/4(2.2014)

                    = 0.5504

antilog 0.5504 = 3.55

(b) $(4.60 \times 10^{20})^{1/5}$ = 1.36×10^4

$\log (4.60 \times 10^{20})^{1/5}$ = 1/5 log (4.60×10^{20})

                         = 1/5 (20.6628)

                         = 4.1326

antilog 4.1326 = $10^{4.1326}$

               = $10^{0.1326}$ x 10^4

               = 1.36 x 10^4

LOGARITHMS AND ANTILOGARITHMS ON THE SLIDE RULE

The L scale on the slide rule is actually a three-place table of
logarithms. All of the operations described in this unit thus far can be
easily carried out using the slide rule rather than log tables. For most
calculations encountered in general chemistry, where a three-place
(or even two-place) table of logarithms is sufficiently accurate, the
L scale of the slide rule is much more convenient for performing these
operations.

The procedure for determining logarithms or antilogarithms with
the slide involves the use of the D and L scales. To find the logarithm
of a number, the characteristic is determined in the usual manner. The
digit sequence for which the mantissa is desired is then located on the

D scale using the hairline of the indicator. The mantissa of the logarithm
of the number is read under the hairline on the L scale. Remember, when
working with numbers less than 1 or greater than 10, that it is more convenient to express the number using standard exponential notation, since
the power of ten is the characteristic of the logarithm.

The procedure for determining antilogarithms is the reverse of the
above procedure, i.e., the mantissa is located on the L scale using the
hairline of the indicator and the number which has this mantissa is read
under the hairline on the D scale. (See Figure 5-1.) Again, logarithms
of very large and very small numbers (as indicated by the characteristic)
are more easily handled if the rules for the manipulation of powers of
ten are observed.

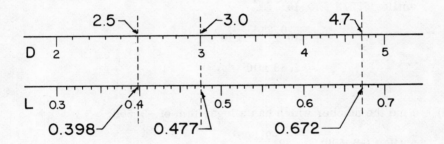

Figure 5-1.   Logarithms of Numbers: log 2.5 = 0.398;

log 3.0 = 0.477; log 4.7 = 0.672.

Examples

(a)   Find the logarithm of 3.80×10^5.

$$\log (3.80 \times 10^5) = \log 3.80 + \log 10^5$$
$$= \log 3.80 + 5$$

$$= 0.580 + 5$$

$$= 5.580$$

(b)   Find the logarithm of 4.25×10^{-7}.

$$\log (4.25 \times 10^{-7}) \quad = \log 4.25 + \log 10^{-7}$$

$$= \log 4.25 - 7$$

$$= 0.628 - 7$$

$$= -6.372$$

(c)   Find the number which has a logarithm of 5.274.

$$\text{antilog } 5.273 \quad = 10^{5.274}$$

$$= 10^{0.274} \times 10^{5}$$

$$= 1.88 \times 10^{5}$$

(d)   Find the number which has a logarithm of -9.500.

$$\text{antilog } (-9.500) \quad = 10^{-9.500}$$

$$= 10^{0.500} \times 10^{-10}$$

$$= 3.16 \times 10^{-10}$$

(e)   Given that, pH $= -\log [H^{+}]$, calculate pH, if $[H^{+}] = 2.7 \times 10^{-4}$.

$$\text{pH} \quad = -\log [H^{+}]$$

$$= -\log (2.7 \times 10^{-4})$$

$$= -(0.43 - 4)$$

$$= 2.57$$

In pH calculations of this type, the mantissa is usually rounded off to two decimal places because pH measurements are not ordinarily determined more precisely than ±0.01 pH unit.

(f)   Calculate pH, if $[H^+] = 5.8 \times 10^{-10}$

$$pH = - \log [H^+]$$

$$= - \log (5.8 \times 10^{-10})$$

$$= - (0.75 - 10)$$

$$= 9.24$$

(g)   Given that, $[H^+] = 10^{-pH}$, calculate $[H^+]$ if pH = 4.74.

(Notice that since logarithms are powers of ten, that this problem simply requires the determination of the antilogarithm of −4.74.)

$$[H^+] = 10^{-pH}$$

$$= 10^{-4.74}$$

$$= 10^{0.26} \times 10^{-5}$$

$$= 1.8 \times 10^{-5}$$

(h)   Calculate $[H^+]$, if pH = 9.7

$$[H^+] = 10^{-pH}$$

$$= 10^{-9.7}$$

$$= 10^{0.3} \times 10^{-10}$$

$$= 2 \times 10^{10}$$

NATURAL LOGARITHMS

Common logarithms or logarithms to the base 10 have been the topic of interest thus far in this unit. There are, however, certain relationships that are encountered in general chemistry in which logarithms to the base $\underline{e}$ are involved, such as:

$$G^o = -RT \log_e K$$

where

$$e = 2.718$$

Logarithms to the base 2.718 are known as natural logarithms and are given the designation "ln" to distinguish them from common logarithms, i.e., "log". The relationship between natural logarithms and common logarithms is given by the equation:

$$\ln a = 2.303 \log a$$

Therefore, the conversion from common logarithms to natural logarithms or vice versa can be easily made using this relationship.

Examples

(a)   Find $\ln (3.60 \times 10^5)$.

$$\begin{aligned}
\ln (3.60 \times 10^5) &= 2.303 \log (3.60 \times 10^5) \\
&= 2.303 (\log 3.60 + \log 10^5) \\
&= 2.303 (0.556 + 5) \\
&= 12.795
\end{aligned}$$

(b) Find ln (5.75×10^{-6})

$$\ln (5.75 \times 10^{-6}) = 2.303 \log (5.75 \times 10^{-6})$$

$$= 2.303 (0.76 - 6)$$

$$= 2.303 (-5.24)$$

$$= -12.07$$

(c) If ln a = 7.385, find log a.

ln a = 2.303 log a

7.385 = 2.303 log a

$\dfrac{7.385}{2.303}$ = log a

3.207 = log a

PROBLEMS

1.   Give the characteristic of the logarithm of the following numbers.

    (a)   5285

    (b)   278

    (c)   0.0045

    (d)   1.92

    (e)   0.000,000,51

    (f) 2.8×10^8

    (g) 4.9×10^{-11}

    (h)   47

    (i)   1,850,000

2.   Give the mantissa of the logarithm of the following numbers.

    (a)   825

    (b)   8.25

    (c)   140

    (d)   0.1752

    (e)   1525

    (f) 2.87×10^{-5}

    (g) 9.87×10^{10}

    (h)   0.6666

3.   Determine the logarithm of each of the numbers in problems 1 and
     2.

4.   Determine the antilogarithm of the following logarithms.

   (a)   1.3201

   (b)   2.4065

   (c)   11.5514

   (d) $\overline{4}$.6484

   (e)   9.6955 – 10

   (f)   4.7300

   (g)   25.9547

   (h) $\overline{16}$.8463

   (i)   –2.5003

   (j)   –9.2

5.   Perform the following operations using logarithms.

   (a)   (275) (85)

   (b)   (1.75) (8.23)

   (c)   (775) (0.172) (9.26)

   (d) $\dfrac{1450}{35}$

   (e) $\dfrac{2850}{193}$

   (f) $\dfrac{(8.55)\ (1980)\ (23.6)}{(0.145)\ (45.7)\ (18.9)}$

6.   Solve the following problems using logarithms.

(a) $(127)^2$

(b) $(17.5)^3$

(c) $(1.85)^9$

(d) $(1450)^{1/2}$

(e) $(128)^{1/3}$

(f) $(985)^{1/5}$

(g) $(0.000,000,725)^{1/4}$

(h) $(2.7 \times 10^3)^9$

(i) $(2.3 \times 10^{-5})^7$

(j) $(8.20 \times 10^9)^{1/3}$

7.   Given that pH $= -\log[H^+]$, determine pH for the following. Slide rule accuracy is sufficient.

(a) $[H^+] = 4.8 \times 10^{-6}$

(b) $[H^+] = 4.57 \times 10^{-6}$

(c) $[H^+] = 4.35 \times 10^{-7}$

(d) $[H^+] = 7.54 \times 10^{-4}$

(e) $[H^+] = 1.75 \times 10^{-9}$

(f) $[H^+] = 9.4 \times 10^{-11}$

8.   Given that $[H^+] = 10^{-pH}$, determine $[H^+]$ for the following. Slide rule accuracy is sufficient.

   (a)   pH = 9.84

   (b)   pH = 2.33

   (c)   pH = 8.45

   (d)   pH = 7.00

   (e)   pH = 10.75

   (f)   pH = 3.55

9.   Solve the following problems. Express all answers in standard exponential notation. Slide rule accuracy is sufficient.

   (a) $x^9 = (4.3 \times 10^{12})$

   (b) $x = (8.3 \times 10^{-6})^{14}$

   (c) $x^{15} = (1.85 \times 10^{-40})$

   (d) $x = (9.4 \times 10^7)^{11}$

   (e) $x^{12} = (2.75 \times 10^{-40})$

   (f) $x^{13} = (4.9 \times 10^{38})$

   (g) $x = (9.42 \times 10^{15})^{11}$

   (h) $x = (8.52 \times 10^{-20})^7$

   (i) $x^9 = (0.000,000,755)$

   (j) $x = (258,000,000)^{14}$

10. Using the relationship, ln a = 2.303 log a, solve the following
     problems. Slide rule accuracy is sufficient.

(a)   Determine ln (4.50×10^{10})

(b)   Determine ln (7.25×10^{-7})

(c)   If ln a = 9.457, determine log a

(d)   If ln a = $\overline{4}.630$, determine log a

Unit 6

SELECTED TOPICS

HAND CALCULATORS

The use of hand calculators by students and educators alike has re-
cently become widespread in colleges and universities throughout the country.
A variety of models are available, marketed by several different companies
currently ranging in price from under $25 to over $500. The less expensive
models can be used only for addition, subtraction, multiplication, and division while the highest price model, in addition to providing the capability
of an expensive slide rule, is equipped with insertable programs for performing certain involved calculations simply. The models in the intermedi-
ate price range ($125-$150) are comparable in versatility to a $2 slide rule.

There are certain advantages in using a hand calculator instead of a
slide rule for computations. The extreme accuracy (i.e., 8-10 digits), and
simplicity of operation are attractive features which hand calculators offer
in comparison to the slide rule. However, before purchasing a $150 hand
calculator the student should keep in mind that 3 digit accuracy is sufficient
for most general chemistry calculations and this is easily obtainable on an
inexpensive slide rule; further, once the operation of the slide rule is learned,
there is little time advantage in using the hand calculator. The added advantage in choosing the slide rule over the hand calculator for computations
in general chemistry is that more <u>thinking</u> is required.

DIRECT PROPORTIONS AND SLOPE OF A LINE

When two quantities are related so that an increase in one causes an
increase in the other, there is a direct relationship between the two quantities. If one of the quantities doubles, triples, etc., when the other quantity

doubles, triples, etc., this direct relationship is a <u>direct proportion</u>.
An example of a direct proportion is the relationship which exists between
the volume of a gas and its absolute temperature at constant pressure,
i.e., if the absolute temperature of the gas doubles, the volume must also
double in order for the pressure to remain constant. The mathematical
relationship involved here is:

$$y = m\,x, \text{ or } y/x = m$$

where m is a <u>proportionality constant</u>

A graphical representation of this relationship is shown in Figure
6-1. Notice that when the two quantities x and y are graphed, a straight
line which passes through the origin results. The constant of proportionality, m, is the <u>slope</u> of the line where,

$$\text{slope} = \frac{y_2 - y_1}{x_2 - x_1} = \frac{\Delta y}{\Delta x}$$

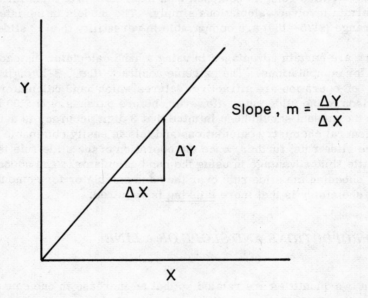

Figure 6-1.

Numerous relationships in the physical sciences where a direct proportionality exists can be conveniently investigated using a graphical approach, e.g., the determination of the proportionality constant, m, from experimental measurements of the quantities x and y.

The general equation for a straight line is,

$$y = m x + b$$

where,   m = slope

b = intercept on the y-axis

Notice that in relationships of direct proportionality, b = 0, i.e., the line intercepts the y-axis at the origin where y = 0. In physical relationships where b ≠ 0, b can be viewed as a "correction factor". As an example, consider the relationship between the Fahrenheit and the Celsius (centigrade) temperature scales. Since both of these scales measure the same physical property in the same manner, it is possible to make a graphical comparison of the two. (See Figure 6-2.)

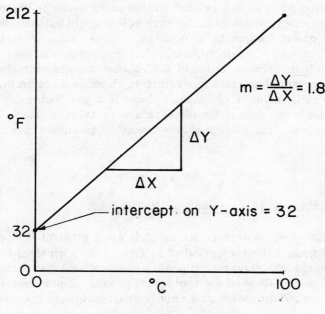

Figure 6-2.

When a comparison is made between two reference points on the two scales, i.e., the freezing point and the boiling point of water, the proportionality constant (slope), and the correction factor (intercept on the y-axis), can be easily determined. The general relationship for a straight line

$$y = m x + b$$

becomes $F = 1.8 C + 32$

where $m = \dfrac{\Delta y}{\Delta x} = 1.8$ (proportionality constant)

$$b = 32 \text{ (correction factor)}$$

The equation is, of course, the familiar relationship used for converting degrees Fahrenheit to degrees Celsius, or vice versa.

INVERSE PROPORTIONS

When two quantities are related so that an increase in one causes a decrease in the other, there is an inverse relationship between the two quantities. If one of the quantities doubles, triples, etc., when the other is reduced to one-half, one-third, etc., this inverse relationship is an inverse proportion. The variation of the volume of a gas with changes in pressure at constant absolute temperature is an example of an inverse proportional relationship, i.e., if the volume of a gas doubles, the pressure is reduced to one-half if the temperature remains constant. For inverse proportions, the relationship between the two quantities, x and y, is

$$y = \frac{m}{x} \quad \text{or} \quad x y = m$$

again, m is a proportionality constant

The graph of an inverse relationship is not a straight line; however, if y is plotted against the reciprocal of x, i.e., $1/x$, a straight line is obtained. The value of the proportionality constant, m, can then be determined from the slope of the reciprocal graph. Experimental measurements of inverse relationships are usually graphed in this manner.

THE QUADRATIC FORMULA

Equations of the type, $ax^2 + bx + c = 0$, are frequently encountered in chemical equilibrium problems. In many cases these problems
can be solved either by extracting the square root of both sides of the
equation, after rearranging, or by making certain simplifying assumptions.
There are, however, some problems which cannot be solved using either
of these approaches, and in cases of this nature it is necessary to apply
the quadratic formula in order to determine the solution. To do this, the
equation must be in the form:

$$ax^2 + bx + c = o$$

where a, b, and c are numbers and x is the unknown. The values of x,
can be determined by substitution into the quadratic formula,

$$x = \sqrt{\frac{-b \pm b^2 - 4ac}{2a}}$$

Notice that two values of x are always obtained when using this relationship. However, when working problems in general chemistry only one
of the roots is the correct answer since the other is physically unrealistic.

Examples

$$x^2 \times (4 \times 10^{-4})x - (4 \times 10^{-5}) = 0$$

Here; $x = 1$

$x = 4 \times 10^{-4}$

$x = -4 \times 10^{-5}$

Substitution into the quadratic formula gives,

$$x = \frac{-4 \times 10^{-4} \pm (4 \times 10^{-4})^2 - 4(-4 \times 10^{-5})}{2}$$

$$x = \frac{-4 \times 10^{-4} \pm \sqrt{16 \times 10^{-8} + 16 \times 10^{-5}}}{2}$$

$$x = \frac{-4 \times 10^{-4} \pm \sqrt{1.6 \times 10^{-4}}}{2}$$

$$x = \frac{-4 \times 10^{-4} \pm 1.26 \times 10^{-2}}{2}$$

$$x = \frac{1.22 \times 10^{-2}}{2} \qquad \text{or} \qquad \frac{-1.30 \times 10^{-2}}{2}$$

$$x = 6.1 \times 10^{-3} \quad \text{or} \quad -6.5 \times 10^{-3}$$

APPENDIX

TABLE OF ATOMIC WEIGHTS

International Atomic Weights—1961 (based on C^{12} = 12 exactly)

Element	Symbol	Atomic number	Atomic weight*	Element	Symbol	Atomic number	Atomic weight*
Actinium	Ac	89	(227)	Mercury	Hg	80	200.59
Aluminum	Al	13	26.9815	Molybdenum	Mo	42	95.94
Americium	Am	95	(243)	Neodymium	Nd	60	144.24
Antimony	Sb	51	121.75	Neon	Ne	10	20.183
Argon	Ar	18	39.948	Neptunium	Np	93	(237)
Arsenic	As	33	74.9216	Nickel	Ni	28	58.71
Astatine	At	85	(210)	Niobium	Nb	41	92.906
Barium	Ba	56	137.34	Nitrogen	N	7	14.0067
Berkelium	Bk	97	(247)	Nobelium	No	102	(254)
Beryllium	Be	4	9.0122	Osmium	Os	76	190.2
Bismuth	Bi	83	208.980	Oxygen	O	8	15.9994
Boron	B	5	10.811	Palladium	Pd	46	106.4
Bromine	Br	35	79.909	Phosphorus	P	15	30.9738
Cadmium	Cd	48	112.40	Platinum	Pt	78	195.09
Calcium	Ca	20	40.08	Plutonium	Pu	94	(244)
Californium	Cf	98	(249)	Polonium	Po	84	(210)
Carbon	C	6	12.01115	Potassium	K	19	39.102
Cerium	Ce	58	140.12	Praseodymium	Pr	59	140.907
Cesium	Cs	55	132.905	Promethium	Pm	61	(145)
Chlorine	Cl	17	35.453	Protactinium	Pa	91	(231)
Chromium	Cr	24	51.996	Radium	Ra	88	(226)
Cobalt	Co	27	58.9332	Radon	Rn	86	(222)
Copper	Cu	29	63.54	Rhenium	Re	75	186.2
Curium	Cm	96	(245)	Rhodium	Rh	45	102.905
Dysprosium	Dy	66	162.50	Rubidium	Rb	37	85.47
Einsteinium	Es	99	(254)	Ruthenium	Ru	44	101.07
Erbium	Er	68	167.26	Samarium	Sm	62	150.35
Europium	Eu	63	151.96	Scandium	Sc	21	44.956
Fermium	Fm	100	(252)	Selenium	Se	34	78.96
Fluorine	F	9	18.9984	Silicon	Si	14	28.086
Francium	Fr	87	(223)	Silver	Ag	47	107.870
Gadolinium	Gd	64	157.25	Sodium	Na	11	22.9898
Gallium	Ga	31	69.72	Strontium	Sr	38	87.62
Germanium	Ge	32	72.59	Sulfur	S	16	32.064
Gold	Au	79	196.967	Tantalum	Ta	73	180.948
Hafnium	Hf	72	178.49	Technetium	Tc	43	(99)
Helium	He	2	4.0026	Tellurium	Te	52	127.60
Holmium	Ho	67	164.930	Terbium	Tb	65	158.924
Hydrogen	H	1	1.00797	Thallium	Tl	81	204.37
Indium	In	49	114.82	Thorium	Th	90	232.038
Iodine	I	53	126.9044	Thulium	Tm	69	168.934
Iridium	Ir	77	192.2	Tin	Sn	50	118.69
Iron	Fe	26	55.847	Titanium	Ti	22	47.90
Krypton	Kr	36	83.80	Tungsten	W	74	183.85
Lanthanum	La	57	138.91	Uranium	U	92	238.03
Lawrencium	Lw	103	(257)	Vanadium	V	23	50.942
Lead	Pb	82	207.19	Xenon	Xe	54	131.30
Lithium	Li	3	6.939	Ytterbium	Yb	70	173.04
Lutetium	Lu	71	174.97	Yttrium	Y	39	88.905
Magnesium	Mg	12	24.312	Zinc	Zn	30	65.37
Manganese	Mn	25	54.9380	Zirconium	Zr	40	91.22
Mendelevium	Md	101	(256)				

*Mass numbers of the most stable known isotopes are given in parentheses.

95

PERIODIC CHART OF THE ELEMENTS

1a	2a	3b	4b	5b	6b	7b	8	8	8	1b	2b	3a	4a	5a	6a	7a	0
1 **H** 1.0080																	2 **He** 4.0026
3 **Li** 6.939	4 **Be** 9.0122											5 **B** 10.811	6 **C** 12.011	7 **N** 14.007	8 **O** 15.999	9 **F** 18.998	10 **Ne** 20.183
11 **Na** 22.990	12 **Mg** 24.312											13 **Al** 26.982	14 **Si** 28.086	15 **P** 30.974	16 **S** 32.064	17 **Cl** 35.453	18 **Ar** 39.948
19 **K** 39.102	20 **Ca** 40.08	21 **Sc** 44.956	22 **Ti** 47.90	23 **V** 50.942	24 **Cr** 51.996	25 **Mn** 54.938	26 **Fe** 55.847	27 **Co** 58.933	28 **Ni** 58.71	29 **Cu** 63.54	30 **Zn** 65.37	31 **Ga** 69.72	32 **Ge** 72.59	33 **As** 74.922	34 **Se** 78.96	35 **Br** 79.909	36 **Kr** 83.80
37 **Rb** 85.47	38 **Sr** 87.62	39 **Y** 88.905	40 **Zr** 91.22	41 **Nb** 92.906	42 **Mo** 95.94	43 **Tc** (99)	44 **Ru** 101.07	45 **Rh** 102.91	46 **Pd** 106.4	47 **Ag** 107.87	48 **Cd** 112.40	49 **In** 114.82	50 **Sn** 118.69	51 **Sb** 121.75	52 **Te** 127.60	53 **I** 126.90	54 **Xe** 131.30
55 **Cs** 132.91	56 **Ba** 137.34	57* **La** 138.91	72 **Hf** 178.49	73 **Ta** 180.95	74 **W** 183.85	75 **Re** 186.2	76 **Os** 190.2	77 **Ir** 192.2	78 **Pt** 195.09	79 **Au** 196.97	80 **Hg** 200.59	81 **Tl** 204.37	82 **Pb** 207.19	83 **Bi** 208.98	84 **Po** (210)	85 **At** (210)	86 **Rn** (222)
87 **Fr** (223)	88 **Ra** (226)	89** **Ac** (227)															

s subshell   d subshell   f subshell   p subshell

*Lanthanum Series

58 **Ce** 140.12	59 **Pr** 140.91	60 **Nd** 144.24	61 **Pm** (147)	62 **Sm** 150.35	63 **Eu** 151.96	64 **Gd** 157.25	65 **Tb** 158.92	66 **Dy** 162.50	67 **Ho** 164.93	68 **Er** 167.26	69 **Tm** 168.93	70 **Yb** 173.04	71 **Lu** 174.97

**Actinium Series

90 **Th** 232.04	91 **Pa** (231)	92 **U** 238.03	93 **Np** (237)	94 **Pu** (242)	95 **Am** (243)	96 **Cm** (247)	97 **Bk** (247)	98 **Cf** (249)	99 **Es** (254)	100 **Fm** (253)	101 **Md** (256)	102 **No** (256)	103 **Lr** (257)

MATHEMATICAL FORMULAS

Arithmetic

Area of a square $= l^2$ (l = length)

Area of a rectangle $= l \times w$ (w = width)

Area of a triangle $= \dfrac{l \times h}{2}$ (h = height)

Circumference of a circle $= 2\pi r = \pi d$ (r = radius, d = diameter)

Area of a circle $= \pi r^2$

Volume of a cube $= l^3$

Volume of a regular prism $=$ area of base $\times$ h

Volume of a pyramid $= 1/3$ area of base $\times$ h

Surface of a sphere $= 4\pi r^2 = \pi d^2$

Volume of a sphere $= 4/3 \pi r^3$

Volume of a cylinder $= \pi r^2 h$

Algebra

$(a+b)^2 = a^2 + 2ab + b^2$

$(a-b)^2 = a^2 - 2ab + b^2$

$(a+b)^3 = a^3 + 3a^2b + 3ab^2 + b^3$

$$(a-b)^3 = a^3 - 3a^2b + 3ab^2 - b^3$$

$$a^x \times a^y = a^{x+y}$$

$$\frac{a^x}{a^y} = a^{x-y}$$

$$a^{-x} = \frac{1}{a^x}$$

$$(a^x)^y = a^{xy}$$

quadratic equations: $ax^2 + bx + c = 0$

$$x = \sqrt{\frac{-b \pm b^2 - 4ac}{2a}}$$

Logarithms

$$\log (a \times b) = \log a + \log b$$

$$\log \left(\frac{a}{b}\right) = \log a - \log b$$

$$\log (a)^n = n \log a$$

$$\log (a)^{1/n} = 1/n \log a$$

$$\ln a = 2.303 \log a$$

$$0.4343 \ln a = \log a$$

Analytical Geometry

Equation of a straight line: $y = mx + b$

$$m = \text{slope} = \frac{y_2 - y_1}{x_2 - x_1} = \frac{\Delta y}{\Delta x}$$

b = intercept on y – axis

TABLE OF FOUR-PLACE LOGARITHMS

N	0	1	2	3	4	5	6	7	8	9	1	2	3	4	5	6	7	8	9
10	0000	0043	0086	0128	0170	0212	0253	0294	0334	0374	4	8	12	17	21	25	29	33	37
11	0414	0453	0492	0531	0569	0607	0645	0682	0719	0755	4	8	11	15	19	23	26	30	34
12	0792	0828	0864	0899	0934	0969	1004	1038	1072	1106	3	7	10	14	17	21	24	28	31
13	1139	1173	1206	1239	1271	1303	1335	1367	1399	1430	3	6	10	13	16	19	23	26	29
14	1461	1492	1523	1553	1584	1614	1644	1673	1703	1732	3	6	9	12	15	18	21	24	27
15	1761	1790	1818	1847	1875	1903	1931	1959	1987	2014	3	6	8	11	14	17	20	22	25
16	2041	2068	2095	2122	2148	2175	2201	2227	2253	2279	3	5	8	11	13	16	18	21	24
17	2304	2330	2355	2380	2405	2430	2455	2480	2504	2529	2	5	7	10	12	15	17	20	22
18	2553	2577	2601	2625	2648	2672	2695	2718	2742	2765	2	5	7	9	12	14	16	19	21
19	2788	2810	2833	2856	2878	2900	2923	2945	2967	2989	2	4	7	9	11	13	16	18	20
20	3010	3032	3054	3075	3096	3118	3139	3160	3181	3201	2	4	6	8	11	13	15	17	19
21	3222	3243	3263	3284	3304	3324	3345	3365	3385	3404	2	4	6	8	10	12	14	16	18
22	3424	3444	3464	3483	3502	3522	3541	3560	3579	3598	2	4	6	8	10	12	14	15	17
23	3617	3636	3655	3674	3692	3711	3729	3747	3766	3784	2	4	6	7	9	11	13	15	17
24	3802	3820	3838	3856	3874	3892	3909	3927	3945	3962	2	4	5	7	9	11	12	14	16
25	3979	3997	4014	4031	4048	4065	4082	4099	4116	4133	2	3	5	7	9	10	12	14	15
26	4150	4166	4183	4200	4216	4232	4249	4265	4281	4298	2	3	5	7	8	10	11	13	15
27	4314	4330	4346	4362	4378	4393	4409	4425	4440	4456	2	3	5	6	8	9	11	13	14
28	4472	4487	4502	4518	4533	4548	4564	4579	4594	4609	2	3	5	6	8	9	11	12	14
29	4624	4639	4654	4669	4683	4698	4713	4728	4742	4757	1	3	4	6	7	9	10	12	13
30	4771	4786	4800	4814	4829	4843	4857	4871	4886	4900	1	3	4	6	7	9	10	11	13
31	4914	4928	4942	4955	4969	4983	4997	5011	5024	5038	1	3	4	6	7	8	10	11	12
32	5051	5065	5079	5092	5105	5119	5132	5145	5159	5172	1	3	4	5	7	8	9	11	12
33	5185	5198	5211	5224	5237	5250	5263	5276	5289	5302	1	3	4	5	6	8	9	10	12
34	5315	5328	5340	5353	5366	5378	5391	5403	5416	5428	1	3	4	5	6	8	9	10	11
35	5441	5453	5465	5478	5490	5502	5514	5527	5539	5551	1	2	4	5	6	7	9	10	11
36	5563	5575	5587	5599	5611	5623	5635	5647	5658	5670	1	2	4	5	6	7	8	10	11
37	5682	5694	5705	5717	5729	5740	5752	5763	5775	5786	1	2	3	5	6	7	8	9	10
38	5798	5809	5821	5832	5843	5855	5866	5877	5888	5899	1	2	3	5	6	7	8	9	10
39	5911	5922	5933	5944	5955	5966	5977	5988	5999	6010	1	2	3	4	5	7	8	9	10
40	6021	6031	6042	6053	6064	6075	6085	6096	6107	6117	1	2	3	4	5	6	8	9	10
41	6128	6138	6149	6160	6170	6180	6191	6201	6212	6222	1	2	3	4	5	6	7	8	9
42	6232	6243	6253	6263	6274	6284	6294	6304	6314	6325	1	2	3	4	5	6	7	8	9
43	6335	6345	6355	6365	6375	6386	6395	6405	6415	6425	1	2	3	4	5	6	7	8	9
44	6435	6444	6454	6464	6474	6484	6493	6503	6513	6522	1	2	3	4	5	6	7	8	9
45	6532	6542	6551	6561	6571	6580	6590	6599	6609	6618	1	2	3	4	5	6	7	8	9
46	6628	6637	6646	6656	6665	6675	6684	6693	6702	6712	1	2	3	4	5	6	7	7	8
47	6721	6730	6739	6749	6758	6767	6776	6785	6794	6803	1	2	3	4	5	5	6	7	8
48	6812	6821	6830	6839	6848	6857	6866	6875	6884	6893	1	2	3	4	4	5	6	7	8
49	6902	6911	6920	6928	6937	6946	6955	6964	6972	6981	1	2	3	4	4	5	6	7	8
50	6990	6998	7007	7016	7024	7033	7042	7050	7059	7067	1	2	3	3	4	5	6	7	8
51	7076	7084	7093	7101	7110	7118	7126	7135	7143	7152	1	2	3	3	4	5	6	7	8
52	7160	7168	7177	7185	7193	7202	7210	7218	7226	7235	1	2	2	3	4	5	6	7	7
53	7243	7251	7259	7267	7275	7284	7292	7300	7308	7316	1	2	2	3	4	5	6	6	7
54	7324	7332	7340	7348	7356	7364	7372	7380	7388	7396	1	2	2	3	4	5	6	6	7
	0	1	2	3	4	5	6	7	8	9	1	2	3	4	5	6	7	8	9

N	0	1	2	3	4	5	6	7	8	9	1	2	3	4	5	6	7	8	9
55	7404	7412	7419	7427	7435	7443	7451	7459	7466	7474	1	2	2	3	4	5	5	6	7
56	7482	7490	7497	7505	7513	7520	7528	7536	7543	7551	1	2	2	3	4	5	5	6	7
57	7559	7566	7574	7582	7589	7597	7604	7612	7619	7627	1	2	2	3	4	5	5	6	7
58	7634	7642	7649	7657	7664	7672	7679	7686	7694	7701	1	1	2	3	4	4	5	6	7
59	7709	7716	7723	7731	7738	7745	7752	7760	7767	7774	1	1	2	3	4	4	5	6	7
60	7782	7789	7796	7803	7810	7818	7825	7832	7839	7846	1	1	2	3	4	4	5	6	6
61	7853	7860	7868	7875	7882	7889	7896	7903	7910	7917	1	1	2	3	4	4	5	6	6
62	7924	7931	7938	7945	7952	7959	7966	7973	7980	7987	1	1	2	3	3	4	5	6	6
63	7992	8000	8007	8014	8021	8028	8035	8041	8048	8055	1	1	2	3	3	4	5	5	6
64	8062	8069	8075	8082	8089	8096	8102	8109	8116	8122	1	1	2	3	3	4	5	5	6
65	8129	8136	8142	8149	8156	8162	8169	8176	8182	8189	1	1	2	3	3	4	5	5	6
66	8195	8202	8209	8215	8222	8228	8235	8241	8248	8254	1	1	2	3	3	4	5	5	6
67	8261	8267	8274	8280	8287	8293	8299	8306	8312	8319	1	1	2	3	3	4	5	5	6
68	8325	8331	8338	8344	8351	8357	8363	8370	8376	8382	1	1	2	3	3	4	4	5	6
69	8388	8395	8401	8407	8414	8420	8426	8432	8439	8445	1	1	2	2	3	4	4	5	6
70	8451	8457	8463	8470	8476	8482	8488	8494	8500	8506	1	1	2	2	3	4	4	5	6
71	8513	8519	8525	8531	8537	8543	8549	8555	8561	8567	1	1	2	2	3	4	4	5	5
72	8573	8579	8585	8591	8597	8603	8609	8615	8621	8627	1	1	2	2	3	4	4	5	5
73	8633	8639	8645	8651	8657	8663	8669	8675	8681	8686	1	1	2	2	3	4	4	5	5
74	8692	8698	8704	8710	8716	8722	8727	8733	8739	8745	1	1	2	2	3	4	4	5	5
75	8751	8756	8762	8768	8774	8779	8785	8791	8797	8802	1	1	2	2	3	3	4	5	5
76	8808	8814	8820	8825	8831	8837	8842	8848	8854	8859	1	1	2	2	3	3	4	5	5
77	8865	8871	8876	8882	8887	8893	8899	8904	8910	8915	1	1	2	2	3	3	4	4	5
78	8921	8927	8932	8938	8943	8949	8954	8960	8965	8971	1	1	2	2	3	3	4	4	5
79	8976	8982	8987	8993	8998	9004	9009	9015	9020	9025	1	1	2	2	3	3	4	4	5
80	9031	9036	9042	9047	9053	9058	9063	9069	9074	9079	1	1	2	2	3	3	4	4	5
81	9085	9090	9096	9101	9106	9112	9117	9122	9128	9133	1	1	2	2	3	3	4	4	5
82	9138	9143	9149	9154	9159	9165	9170	9175	9180	9186	1	1	2	2	3	3	4	4	5
83	9191	9196	9201	9206	9212	9217	9222	9227	9232	9238	1	1	2	2	3	3	4	4	5
84	9243	9248	9253	9258	9263	9269	9274	9279	9284	9289	1	1	2	2	3	3	4	4	5
85	9294	9299	9304	9309	9315	9320	9325	9330	9335	9340	1	1	2	2	3	3	4	4	5
86	9345	9350	9355	9360	9365	9370	9375	9380	9385	9390	1	1	2	2	3	3	4	4	5
87	9395	9400	9405	9410	9415	9420	9425	9430	9435	9440	0	1	1	2	2	3	3	4	4
88	9445	9450	9455	9460	9465	9469	9474	9479	9484	9489	0	1	1	2	2	3	3	4	4
89	9494	9499	9504	9509	9513	9518	9523	9528	9533	9538	0	1	1	2	2	3	3	4	4
90	9542	9547	9552	9557	9562	9566	9571	9576	9581	9586	0	1	1	2	2	3	3	4	4
91	9590	9595	9600	9605	9609	9614	9619	9624	9628	9633	0	1	1	2	2	3	3	4	4
92	9638	9643	9647	9652	9657	9661	9666	9671	9675	9680	0	1	1	2	2	3	3	4	4
93	9685	9689	9694	9699	9703	9708	9713	9717	9722	9727	0	1	1	2	2	3	3	4	4
94	9731	9736	9741	9745	9750	9754	9759	9763	9768	9773	0	1	1	2	2	3	3	4	4
95	9777	9782	9786	9791	9795	9800	9805	9809	9814	9818	0	1	1	2	2	3	3	4	4
96	9823	9827	9832	9836	9841	9845	9850	9854	9859	9863	0	1	1	2	2	3	3	4	4
97	9868	9872	9877	9881	9886	9890	9894	9899	9903	9908	0	1	1	2	2	3	3	4	4
98	9912	9917	9921	9926	9930	9934	9939	9943	9948	9952	0	1	1	2	2	3	3	4	4
99	9956	9961	9965	9969	9974	9978	9983	9987	9991	9996	0	1	1	2	2	3	3	3	4
	0	1	2	3	4	5	6	7	8	9	1	2	3	4	5	6	7	8	9

ANSWERS TO PROBLEMS

UNIT 1   (Exponents: Powers of Ten)

1. a. 7.2×10^{6} j. 2.73×10^{-1}

     b. 2.24×10^{4} k. 2.1×10^{-7}

     c. 2.24×10^{1} l. 1.06×10^{1}

     d. 8×10^{3} m. 8.30×10^{6}

     e. 1.25×10^{5} n. 7.2×10^{-4}

     f. 1.4×10^{0} o. 3.5×10^{-4}

     g. 7×10^{-3} p. 1.973×10^{-2}

     h. 3.5×10^{-4} q. 2×10^{2}

     i. 9.05×10^{-3} r. 5×10^{1}

2. a. 0.007

     b. 210,000

     c. 1.4

     d. 0.000505

     e. 0.000,000,000,93

     f. 600,000,000,000,000,000,000,000

     g. 0.000,000,000,000,000,000,000,0017

     h. 21

3. a. 6.1×10^4

    b. 4.5×10^{-3}

    c. $14.3 \times 10^2 = 1.43 \times 10^3$

    d. 4.7×10^{-2}

    e. $0.5 \times 10^5 = 5 \times 10^4$

    f. 8.41×10^5

    g. 1.61×10^{-7}

    h. 9.63×10^4

    i. 9.00×10^{-10}

    j. 2.65×10^{-2}

    k. 4.56×10^{-2}

    l. 7.71×10^{-2}

4. (a) 1.2×10^6

    (b) 1.92×10^1

    (c) 3.36×10^{-3}

    (d) 4.5×10^{10}

    (e) 1.568×10^{-6}

    (f) 3.9×10^{-2}

    (g) 2.47×10^{38}

    (h) 7.2×10^{12}

    (i) 6×10^{-9}

    (j) 5.1×10^{-7}

    (k) 1.59×10^9

    (l) 2.75×10^{-14}

    (m) 4.57×10^{18}

    (n) 1.55×10^{-10}

    (o) 4.78×10^6

5. (a) 1.5×10^{1} (h) 1.2×10^{-22}

    (b) 3×10^{2} (i) 6.3×10^{28}

    (c) 3.52×10^{-2} (j) 2.0×10^{5}

    (d) 3.0×10^{-2} (k) 5.4×10^{-15}

    (e) 5×10^{0} (l) 2.0×10^{2}

    (f) 6.5×10^{-10} (m) 4.99×10^{25}

    (g) 2.00×10^{-2} (n) 1.1×10^{1}

6. (a) 5.0×10^{-9} (f) 4.2×10^{7}

    (b) 2.0×10^{10} (g) 1.02×10^{8}

    (c) 1.82×10^{10} (h) 5.0×10^{15}

    (d) 1.3×10^{9} (i) 1.13×10^{3}

    (e) 9.5×10^{-3}

7. (a) 4.0×10^{7} (i) 4.48×10^{0}

    (b) 4.25×10^{12} (j) 7.68×10^{-1}

    (c) 5.8×10^{8} (k) 4.2×10^{-1}

    (d) 1.62×10^{13} (l) 2.5×10^{5}

    (e) 4.5×10^{-7} (m) 1.3×10^{0}

    (f) 1.36×10^{-12} (n) 9.2×10^{-2}

    (g) 6×10^{-10} (o) 8.2×10^{21}

    (h) 3.7×10^{-20} (p) 8.3×10^{-28}

8.   (a) 4×10^8

     (b) 1.68×10^7

     (c) 3.97×10^{-3}

     (d) 1.44×10^{-16}

     (e) 8.84×10^{19}

     (f) 2.74×10^6

     (g) 6.86×10^{-6}

     (h) 1.38×10^{-14}

     (i) 6.4×10^{43}

     (j) 8.0×10^{-60}

     (k) 2.56×10^{-6}

     (l) 1.97×10^{-5}

     (m) 1.25×10^{14}

     (n) 2.5×10^{11}

     (o) 3.6×10^{-13}

     (p) 2.7×10^{-23}

     (q) 4.3×10^{-10}

     (r) 1.4×10^{10}

     (s) 1.2×10^{18}

     (t) 1.7×10^{-13}

     (u) 4×10^{-2}

     (v) 1×10^7

     (w) 1.1×10^{-7}

     (x) 5×10^{-2}

     (y) 9×10^5

     (z) 4×10^2

9.   (a) 1.6×10^{15}

     (b) 1.49×10^1

     (c) 4.8×10^4

     (d) 4×10^0

     (e) 9.0×10^5

     (f) 6.9×10^{-17}

     (g) 5×10^2

     (h) 4×10^{-3}

UNIT 2   (Significant Figures)

1. (a) 3   (f) 3   (k) 6

    (b) 2   (g) 1   (l) 3

    (c) 3   (h) 4   (m) 1

    (d) 2   (i) 7   (n) 1

    (e) 5   (j) 5

2. (a) 181.9   (m) 1.2×10^{18}

    (b) 334.57   (n) 4×10^{-6}

    (c) 4.3590   (o) 8.8×10^{19}

    (d) 8.10×10^{5} (p) 1.6×10^{-14}

    (e) 1.73×10^{-7} (q) 2.2×10^{-8}

    (f) 4.5×10^{10} (r) 2.3×10^{8}

    (g) 8.3×10^{5} (s) 4.674

    (h) 8.6×10^{-12} (t) 5.6893

    (i) 5.83×10^{20} (u) 3.00×10^{3}

    (j) 2.1×10^{-6} (v) 4.0×10^{-4}

    (k) 4.5×10^{3} (w) 4.0×10^{3}

    (l) 1.1×10^{-19} (x) 3.0×10^{-3}

UNIT 3   (The Metric System; Conversion Factors)

1.   (a)   1.5 m

(b) 5.2×10^3 g

(c) 2×10^{-3} l

(d)   750 mm

(e) 5.40×10^{-4} km

(f) 5.0×10^6 cg

(g) 4.5×10^{-3} kg

(h) 1.4×10^7 μg

(i)   121 km

(j) 8.2×10^{-10} ft

(k) 7.6×10^4 μ

(l) 1.5×10^{-6} cm

(m) 1.14×10^{16} pg

(n) 1.93×10^{-8} Tm

(o)   4.73 l

(p) 7.57×10^4 ml

(q) 1.99×10^7 μl

(r)   4.47 m/sec

(s) 5×10^{-1} nm

(t)   1.5 l

(u) 2.4×10^{-3} Gsec

(v) 4.4×10^{-5} ton

(w) 7×10^4 oz

(x) 5.76×10^3 in^2

(y) 2.5×10^4 mm^3

(z) 3.17×10^{-4} qt

2. 2.9×10^3 cm/sec

3.   967.5 cm^2

4.   216 cm^3

5.   655.5 cm^3

6. 2.5×10^{-15} cm^2

7. 6.6×10^8 fathoms/forthnight

8.   7.34 firkins

9.   (a)   27.5 mm

        (b)   1.08 in

        (c) 2.75×10^{8} A

        (d) 3.01×10^{-2} yd

10.   (a) 1.6×10^{-24} in^3

        (b) 2.7×10^{-23} ml

        (c) 9.5×10^{-28} ft^3

        (d) 2.7×10^{-26} l

UNIT 4   (The Slide Rule)

1.   (a) $420 = 4.2 \times 10^2$

      (b) $14.7 = 1.5 \times 10^1$

      (c) $40.32 = 4.0 \times 10^1$

      (d) $20.75 = 2.1 \times 10^1$

      (e) $160.6 = 1.6 \times 10^2$

      (f) $3.64 \times 10^8 = 3.6 \times 10^8$

      (g) $5.94 \times 10^{-9} = 5.9 \times 10^{-9}$

      (h) $13.65 \times 10^8 = 1.4 \times 10^9$

      (i) $26.66 \times 10^{-5} = 2.7 \times 10^{-4}$

      (j) $6.54 = 6.5$

      (k) $50 = 5.0 \times 10^1$

      (l) $282.9 = 2.83 \times 10^2$

      (m) $1.65 \times 10^2 = 1.6 \times 10^2$

      (n) $0.20 \times 10^{-1} = 2.0 \times 10^{-2}$

      (o) $1.47 \times 10^{45} = 1.5 \times 10^{45}$

      (p) $137.5 = 1.4 \times 10^2$

      (q) 4.5×10^{-1}

      (r) $1.46 \times 10^{-10} = 1.5 \times 10^{-10}$

      (s) $2.64 \times 10^7 = 2.6 \times 10^7$

2.   (a) $1412 = 1.4 \times 10^3$

      (b) $35.45 = 3.5 \times 10^1$

      (c) $0.68 \times 10^4 = 6.8 \times 10^3$

      (d) $1.45 \times 10^{-11} = 1.4 \times 10^{-11}$

      (e) $7.66 \times 10^9 = 7.7 \times 10^9$

      (f) $10.24 \times 10^6 = 1.0 \times 10^7$

      (g) $0.272 \times 10^{-21} = 2.7 \times 10^{-22}$

3.   (a)   5.52

      (b)   9.6

      (c)   27

      (d)   4.1

      (e) $42.87 = 43$

      (f) $91.1 \times 10^3 = 9.1 \times 10^4$

      (g) $23 \times 10^{-12} = 2.3 \times 10^{-11}$

      (h) $15.62 \times 10^{-15} = 1.6 \times 10^{-14}$

      (i) 3.42×10^{10}

      (j) 3.6×10^{14}

      (k) $614 \times 10^{12} = 6.14 \times 10^{14}$

      (l) 1.86×10^{27}

3.   (m)   4.06

(n)   1.7

(o)   13.5

(p) $2.19 \times 10^{-3} = 2.2 \times 10^{-3}$

(q) 9.65×10^{3}

(r)   1.20

(s)   2.60

(t)   5.59

(u) 6.59×10^{1}

(v) $1.42 \times 10^{-3} = 1.4 \times 10^{-3}$

4.   (a) $112 \times 10^{-2} = 1.1 \times 10^{0}$

(b) $26 \times 10^{10} = 2.6 \times 10^{11}$

(c) $1.48 \times 10^{-5} = 1.5 \times 10^{-5}$

(d) $0.89 \times 10^{-1} = 8.9 \times 10^{-2}$

(e) 5.1×10^{7}

(f) $7.64 \times 10^{2} = 7.6 \times 10^{2}$

(g) 3.60×10^{9}

(h) 2.4×10^{-11}

(i) 4.1×10^{0}

(j) $7.75 \times 10^{-1} = 7.8 \times 10^{-1}$

(k) 1.4×10^{-13}

(l) $71.9 \times 10^{-6} = 7.2 \times 10^{-5}$

(m) $1.65 \times 10^{-7} = 1.6 \times 10^{-7}$

(n) $5.36 \times 10^{45} = 5.4 \times 10^{45}$

(o) $2.66 = 2.7 \times 10^{0}$

UNIT 5 (Logarithms)

1. (a) 3   (d) 0   (g) $\overline{11}$

    (b) 2   (e) $\overline{7}$ (h) 1

    (c) $\overline{3}$ (f) 8   (i) 6

2. (a) 0.9165   (d) 0.2435   (g) 0.9943

    (b) 0.9165   (e) 0.1832   (h) 0.8239

    (c) 0.1461   (f) 0.4579

3. 1- (a) 3.7230   (d) 0.2833   (g) $\overline{11}.6902$

    (b) 2.4440   (e) $\overline{7}.7076$ (h) 1.6721

    (c) $\overline{3}.6532$ (f) 8.4472   (i) 6.2672

    2- (a) 2.9165   (d) $\overline{1}.2435$ (g) 10.9943

    (b) 0.9165   (e) 3.1832   (h) $\overline{1}.8239$

    (c) 2.1461   (f) $\overline{5}.4579$

4. (a) 20.9   (f) 5.37×10^{4}

    (b) 255   (g) 9.01×10^{25}

    (c) 3.56×10^{11} (h) 7.02×10^{-16}

    (d) 4.45×10^{-4} (i) 3.16×10^{-3}

    (e) 4.96×10^{-1} (j) 6.31×10^{-10}

5. (a) $23,375 = 2.3 \times 10^4$ (d) $41.43 = 4.1 \times 10^1$

    (b) $14.402 = 1.4 \times 10^1$ (e) $14.77 = 1.48 \times 10^1$

    (c) $1234.4 = 1.23 \times 10^3$ (f) $3190 = 3.19 \times 10^3$

6. (a) $16,129 = 1.61 \times 10^4$ (f) 3.97

    (b) $5359.4 = 5.36 \times 10^3$ (g) 2.92×10^{-2}

    (c) $253.83 = 2.54 \times 10^2$ (h) 7.6×10^{30}

    (d) $38.08 = 3.808 \times 10^1$ (i) 3.4×10^{-33}

    (e) 5.04 (j) $2.017 \times 10^3 = 2.02 \times 10^3$

7. (a) 5.32 (d) 3.12

    (b) 5.34 (e) 8.76

    (c) 6.36 (f) 10.03

8. (a) 1.4×10^{-10} (d) 1.0×10^{-7}

    (b) 4.7×10^{-3} (e) 1.8×10^{-11}

    (c) 3.6×10^{-9} (f) 2.8×10^{-4}

9. (a) 2.5×10^1 (f) 9.4×10^2

    (b) 1.0×10^{-71} (g) 6.3×10^{175}

    (c) 2.25×10^{-3} (h) 3.2×10^{-134}

    (d) 5.0×10^{87} (i) 2.1×10^{-1}

    (e) 5.0×10^{-4} (j) 6.3×10^{117}

10.   (a)   24.534   (c)   4.106

      (b)   -14.140   (d)   -1.463

1255